ACKNOWLEDGEMENTS

Our sincere and grateful appreciation goes out to those people working in the organizations whose considerate help made this possible. Our special thanks to Ron Miller and Jim Barnett along with their coworkers. And to the many employees of the companies, publishing houses, printers, home owners and photographers, who are far too many to name, our kindest feelings also go–thank you:

ADAMS COUNTY CHAMBER OF COMMERCE
ANDREW FISHER
THE BERGERONS
CALLON PETROLEUM
COLUMBIA PICTURES
DANNY RICHARDSON & HIS FAMILY
DENISE FIELDS & HER FAMILY
DEPOSIT GUARANTY NATIONAL BANK
ED PRINCE
THE GARDEN CLUBS
HISTORIC NATCHEZ FUNDATION:
Christine Foster, Ronald W. Miller and Mabel R. Tippens.
JAMES ENOS
JUDGE GEORGE W. ARMSTRONG LIBRARY
KAMINER & WELCH
MARK COFFEE
McNAUGHTON & GUNN
MISSISSIPPI DEPARTMENT OF ARCHIVES & HISTORY
MISSISSIPPI STATE HIGHWAY DEPT.
NATCHEZ CONVENTION CENTER:
Gladys, Carol Campbell, Lofton Campbell, Shirley Campbell, Jennifer Essary, Jack Gebhardt, Olivia Harris, Lena Tate and Charlie Woods.
NATCHEZ PHOTO LAB:
Larry & Ruth Fisher
NATCHEZ TRACE PARKWAY
NATIONAL PARK SERVICE
JIM NUTTER
RAMADA HILLTOP
RAY BARD
RIDGWAY INC.
SOUTHERN EXPOSURE
TOTAL PHOTOGRAPHICS
UNITED MISSISSIPPI BANK
UNITED STATES ARMY MILITARY HISTORY INSTITUTE
LISA & ARIANE

Antebellum Home "LONGWOOD"

NATCHEZ
The City In History

LEE DABNEY SWINNY

Cover Art by Kirby Brooks Meng
Editor-In-Chief, Mizette Ariane Hinson

AVAVA BOOKS
P.O.Box 120 Natchez, MS 39121
ISBN 0-924043-00-8
Manufactured in the USA

CONTENTS

Introduction

Section I: OLD NATCHEZ

1. **The History Of Natchez 9**
   From the Ice Age through the Civil War
2. **Antebellum Natchez 33**
   Descriptions and Information on the Antebellum Homes

Section II: MODERN NATCHEZ

3. **The Community 57**
   Map of Natchez
   Statistics, Resources, Services & Institutions
4. **Seeing Natchez 65**
   Transportation, Tours, Lodging, Bed & Breakfast,
   Points of Interest, The Nature of Natchez
5. **Walk-A-Bout-Natchez 83**
   Walking guide with Map of downtown Antebellum Homes
6. **Recreation 87**
   Daytime Activities (Shopping, Sports, Hobbies)
7. **Entertainment 99**
   Nightlife (Confederate Pageant, Dining & Dancing,
   Restaurants, Lounges, Movies, and Theatres)
8. **Clendar Of Events 107**
   Annual Events, Scheduled Conventions & Activities

Section III: FUTURE NATCHEZ

9. **The Future Of Natchez & The State Of Mississippi 117**
   Points of Interest close to Natchez,
   Travel Distance to other towns, Map of the State,
   Natchez National Historical Park
   Developments for the Future of Natchez

Index 124

Introduction

Among the times and ages of past and present, with their events and conditions so integral to the development of our future, Natchez stands as an important part of this great nation's past. While the pioneers carved out a piece of heritage for their descendents and struggled hard to build a meanigful society and a future for their young, Natchez grew out of the essence of that desire, to become that heritage herself. Natchez is that heritage. The proof is portayed in the majestic architecture they left here and in the descendents which developed out of those pioneers. All this remains intact in a society which is now the future of our ancestor's dreams. It is left only for todays society to preserve and protect this heritage which makes up all of our lives.

Natchez is The City In History because it is one of the oldest colonies on the Mississippi River. Yet, it is also young because it is part of a young Country. No matter in what age they developed, no matter how many important events took place in them, the towns in this Free Land and the people who built them are all as much a part of history as life today will be thousands of years from now. Along with all of the other towns, communities and cities spread across this Nation, Natchez Is History.

This work is respectfully dedicated
to the women and men like Danny Richardson,
who took risks to preserve our heritage in photographs
like the one on the back of this guide.
And also to my family, who gave me love...

I

OLD NATCHEZ

John J. Audubon painting of Natchez when he lived here. Courtesy Historic Natchez Foundation.

1

The History Of Natchez

Some people come to Natchez and the first thing they do is compare it to other places. Granted, the history of the Nation is similar in many places, as the movements of people and times progress, and Natchez is likewise similar. Yet, Natchez has, over the past two centuries, been a gathering place for uncommon events and people. Because of its rich soil, the Natchez area attracted an overabundance of wealthy cotton planters before the Civil War. When oil was the boom, this industry also flocked in to reap the rewards so generously endowed by nature. Also, over the years, many great names have found their way to Natchez to enjoy its lavish beauty. There have been great writers like Mark Twain, great statesmen like Winthrop Sargent, David Holmes, Henry Clay, Jefferson Davis, Bob Dole and Marquis de Lafayette. There have been many great Presidents like Andrew Jackson, Ulysses S. Grant, William Howard Taft, Zachary Taylor and George Bush; the families of presidents like Johnson, Roosevelt and Madison. Also to visit are artist, actors and professionals like Clarke Gable, Vivien Leigh, John J. Audubon, Jenny Lind, Helen Hayes, George Hamilton, Kris Kristopherson, Muhamed Ali, Andrew Ellicot, Tom Thumb, Ralph Waite, Glen Campbell, Montgomery Clift, Elizabeth Taylor, Edwin Booth, Harvey Corman, Patrick Swayze and Tyrone Power. And there have been famous generals such as Robert E. Lee, John Anthony Quitman and Douglas MacArthur. The list goes on and they often keep coming back to experience what is unique to Natchez–its history.

In fact, visitors and tourists come from all over the world to see what remains untouched of the Antebellum culture that was the Old South. In no other palace can you find such a diverse collection of history, bestowed on Natchez from the people of Spain, France, England, Africa and throughout Europe and the Orient. In no other place is this Nation's history so well preserved and her heritage so honored. Yes, Natchez is a small place compared to the great Metropolis, but the traces left here over the ages show a greatness in its own right.

PRE HISTORY

During the ice age a glacier of tremendous proportions carved the entire Mississippi Valley out of the face of the earth. Over a great period of time, the Loess Bluff was formed by high winds, prehistoric dust storms that blew the surface soils from the west into the great ridge along much of the eastern shore of the Mississippi River. This Ridge was built up in layers varying from thirty to ninety feet in depth, of the soft, loamy soil which contains a thin but very rich topsoil. Rising out of the western, Delta and the River Lowlands areas to the east, and ranging from five to fifteen miles in width, the bluff forms the thin line of hills or highlands that separates these areas from the Brown-Loam area, from one end of the state to the other.

It wasn't until the glaciers melted that the Mississippi River was formed from the run-off of melting ice. As time goes on, the river slowly etches its way into the side of Loess Bluff, causing great chunks of the hillside to plummet into the mighty Mississippi, consequently leaving the tall, vertical, dirt exposures. We call these cliffs The Bluff.

UNDER FIVE FLAGS

Over the growth of this nation, the history of Natchez falls under the ruling flags of five great battling governments, all of which claimed this ground as their own and gave their lives in defense of it. Yet first to lay claim on this land was the American Indian. Many centuries ago he was the lone pioneer–without competition. Maybe this is why he didn't erect a flag such as the white man did. Maybe he didn't think he needed to.

NATCHEZ INDIANS

The extinct Natchez Indians were not the first to settle this area, though it was the American Indian, out of which the Natchez developed, who did come here originally like most of North and South America. It is hard to conceive, but hard evidence proves that the Indian was here more than 11,000 years ago! The Natchez did not migrate here, but *developed* their uniqueness on these same grounds.

The peculiarities of the Natchez, probably picked up from northern Indians (like the larger mound centers, the different style of pottery, their social class structure and their sacrificial rites), did not appear until around A.D. 1200.

All of these great expanses of time frames are hard to conceived as people moved around from area to area, slowly shifting in location and cultural behavior. Looking at a map and trying to picture 11,000 years of movement and development will make a person's vision blur. Yet it is a wonder and a mystery that all of it was even possible.

One of the greatest mysteries is the Emerald Mound. Though the Natchez reached the pinnacle of their culture in the Mid-1500's, Indian mounds were a relatively new thing to them. The forefathers of the Natchez erected the

Transport of the Great Sun of the Natchez. Courtesy Mississippi Dept. of Archives & History.

massive Emerald Mound on the Natchez Trace Parkway, just outside of town, in 1400 A.D. Second in size only to the Nanih Waiya Mound of the Choctaw, the Emerald ceremonial mound covers nearly eight acres. For some reason, they abandoned this engineering giant to end up at the comparatively small Grand Village. It is possible that a new leader was followed to the Grand Village, and the old leaders were abandoned at Emerald Mound, to die and disappear.

Whatever happened, their strange social structure began to deteriorate as it did with other Indian Nations. Though the Natchez were one of the last tribes to hang on to the old way, it and the sacrificial rites began to succumb to more practical forms of Government. There was also the influence of the white man that altered their attitudes. These, along with the new viruses brought by de Soto's army, which killed off many tribes and scattered them out, served to break up the American Indian's pure form of life. By 1700 there were only 30,000 Indians in what is now Mississippi: 5,000-10,000 peaceful Choctaws, 2,000 roaming, warlike Chickasaws, and 3,500 Natchez who were not warlike unless provoked.

The similarities with ours in the Indians' religious beliefs indicate an Oriental origin. The Natchez taught of a great prehistoric flood that covered all the land. The Chickasaw believed in "A Beloved One Who Dwells In The Blue Sky." The Choctaw spoke of the Great Spirit. And the "Happy Hunting Ground" was a belief held by all three tribes.

The Natchez were Sun Worshippers who believed the sun was the Sun God or Great Spirit who created man by kneading earth and water. When the time came, the Great Spirit sent his Son to establish government and to teach religious customs. Then the Son changed himself into a stone. This stone was kept in a temple on a ten-foot mound where an eternal fire burned. This was so significant that the keepers of the fire were put to death if it went out.

The eternal fire symbolized the Sun from which the chief or "Great Sun" descended. The Great Sun's role was to greet the Sun every day with a howl and guide it across the sky with a motion of his arms. When the Great Sun died, his whole family was ceremoniously strangled so they could accompany him into the next life, and their remains were placed in the temple. On its own mound was the Sun's house, which was burned. The mound was raised higher and a new house was built for the New Great Sun, who was the eldest son of the nearest female relative on his maternal side. The Great Sun's own children had no chance of becoming rulers.

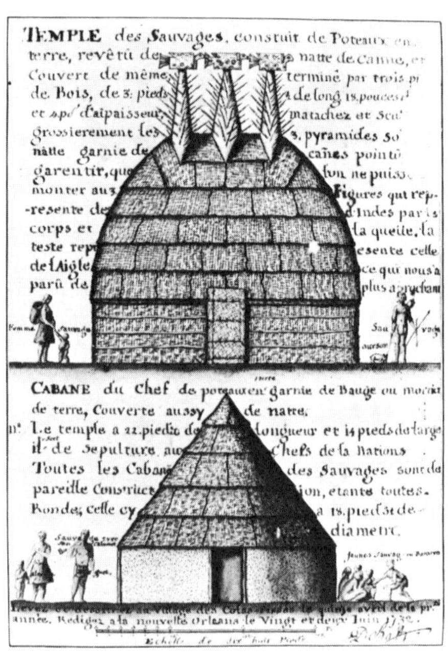

Temple & Hut of the Natchez Indians.
Courtesy Mississippi Dept. of Archives & History.

Because of this female lineage, Mississippi was the first state to establish legal property rights for women. Betty Allen, the Chickasaw wife of Major John L. Allen, was given a slave by her Indian father. When her husband died, his creditors tried to seize the slave. The state's highest court ruled in Betty Allen's favor, basing their decision on the Chickasaw law of inheritance.

Under the Great Sun were two classes of people: the upper class, which consisted of Suns (chief's relatives), Nobles, and the Honored People; and the lower class they called "Stinkards" but never to their face. It was also tribal law that an upper class person had to marry a Stinkard in order to keep the social classes from becoming too fixed.

Even though the female lineage decided inheritance and royal hierarchy, the men led the leisurely lifestyle. The women and children were responsible for most of the work, while the men played games like "Chunky" (similar to handball) and gambled and only helped during planting and harvesting seasons. The husband did hunt, but brought only the animal's ear home for his wife to take into the woods, match with the deer, bear or bison, and lug the carcass back.

Of all the Indians, the Natchez were the most colorful. An attractive people, they were tall, robust, and strong with a "proud air" about them. It was the group of Natchez tribesmen, that was taken to France as an exhibition, that started the idea of the "Noble Savage" when they deceived the Europeans into thinking they were uncorrupted by the habits of civilized men. Later, the French discovered their true nature. However, they were good farmers, paying particular attention to their crops of maize, pumpkin, chestnuts, watermelon and tobacco. They even mixed their tobacco with sumac leaves to create a "blend" as modern tobacconists boast.

Unlike other tribes, the Natchez lived in low, square mud huts with round thatched roofs and without windows. They were colorful, yet they practiced head flattening, ear slitting to make very large holes, blackened their teeth each morning by rubbing a mixture of tobacco and wood ashes, blackened their skin by repeated rubbings of bear's oil (probably for protection against sunburn and mosquitoes), and the men habitually plucked the hair from their scalp except for a small spot on top. The warriors were set apart by tatoos on their right shoulder of the sign of the conquered nation, which they received after a victorious battle. The warriors also drank "Black Drink" before going into battle. It was a brew distilled of Yapon berries or "Texas Holly."

While a close comparison to civilized man showed significant differences, a similarity in other aspects of belief emerged: the Indians' calendar had thirteen months and it began in the "Deer Month" of March as ours once did. Many of the Europeans found that the Indian medicine was better where wounds were concerned and actually went to the medicine man instead of their own doctors. However, some of the treatments came across as a bit superstitious, such as an onion for rattlesnake bite because the onion had "a rattler," and hanging moss for that "sluggish feeling." Even though some of their customs seemed strange, some were similar and few so endearing as the name the Indian gave our great Mississippi River– "The River Of The Holy Spirit."

Ceremonial Dance of the Natchez.
Courtesy Mississippi Dept. of Archives & History.

THE EXPLORERS

DE SOTO'S SEARCH FOR GOLDEN CITIES. It is probable that Spain sent explorers close to the mouth of the Mississippi River before Ponce de Leon ever ventured into the Florida peninsula in 1513, but it was Hernando de Soto, hacking his way from Tampa, Florida through most of the southern states in search of "Golden Cities" and a "Passage to the Orient," who in May of 1541 was the first civilized man to walk up to the great Mississippi River some miles below Memphis. He called it the "Rio Grande de Florida," the great river of Florida. In his desperate search for gold, de Soto used the Indians he found as bearers and in any other way he could only to receive fierce struggle. Battle after bloody battle only kept him away from travelling through Natchez, but upon his death in 1542, the remainder de Soto's ragged men might have passed through here on their way out of the Mississippi Valley. There are even stories that the Natchez Indians chased them out.

MARQUETTE & JOLIET. While Spain explored from the south, France came from the northern region of New France or Canada. Wanting to know of the mysteries of the great river, talked of by other missionaries, the Governor of Canada sent (on May 17, 1673) Father Jaques Marquette and Louis Joliet to find out where the Mississippi went to. In Arkansas the Indians told the explorers that the river went to the Gulf and that the English had supplied the Indians down river with guns. Marquette and Joliet went back to Canada with their discovery.

LA SALLE & TONTY. In August of 1681 La Salle and Tonty set out from Canada to claim and name Louisiana and the Mississippi "The River Colbert" in honor of the great minister who planned the project as part of Louis XIV's grandiose imperial schemes. Before continuing on to the Gulf of Mexico, La Salle and Tonty discovered the Natchez Indians in 1682 and established good relations with them. Upon returning north La Salle built Fort St. Louis on the Illinois River, now Illinois.

Tonty stayed while La Salle returned to France. In 1684 La Salle intended to return from the south and make Louisiana a permanent settlement but, as fate would have it, his expedition was lost and landed at Texas. His sea captain left him with one small boat which sank. La Salle and his remaining men set out across Texas toward Fort St. Louis but, because of their frustration with continued failure and the fear of being lost in the wilderness forever, La Salle's men murdered by him in 1687. Tonty searched the river for La Salle without any recourse but to leave a letter for him with the Natchez Indians.

IBERVILLE & BIENVILLE. Iberville left France with his young brother Bienville as a midshipman on one of two armed ships of families, farmers, mechanics and a carpenter named Penicaut. The Spanish would not allow them near Pensacola so they proceeded to colonize Biloxi. Iberville discovered and named Ship Island (for its silhouette) and Cat Island, named so because its many raccoons were mistaken for cats.

When he anchored at Biloxi, the indians fled. The two sick tribesmen, who were left behind, were given presents. The Indians returned and Bienville remained on shore as a hostage while four tribesmen were taken to the ship. In this was the establishment of good relations with the Biloxis.

Then, by 1699 Iberville and Bienville came up the Mississippi some 300 miles and obtained a coat of mail armor left by de Soto, a letter left by Tonty, and a "speaking bark," or prayer book. Iberville explored lake Ponchatrain and Ocean Spring's Back Bay area, where he built Fort Maurepas, then returned to France with his report. Bienville stayed and named Mobile Bay, Bay St. Louis and met an English ship at the mouth of the Mississippi. Either from the lack of wind caused by the river valley, deceptive oratory or threats of violence by Bienville, the English ship was diverted and to this day that spot is called the English Turn. Bienville also stopped at the mouth of and named the Pearl River, for its banks were covered with the pearls from the shells the Indians used to scrape their canoes. Penicaut the carpenter wrote, "we never after heard of these pearls."

FORT ROSALIE PLANNED. Iberville returned to Biloxi in 1700 to breed buffalo, seek pearls and mines, examine mulberry trees for silk and find timber for ships. Establishing peaceful relations with the Natchez Indians was enough that they granted to France the grounds for Fort Rosalie (La Ville de Rosalie aux Natchez), named after the countess, wife of French Marine Minister Ponchatrain. Whenever Iberville returned, he brought paint and gifts to the Natchez Indians. On one of his trips he met the missionary, Father Davion, ministering to the Tunicas south of Natchez at what was called Davion's Rock, now called Fort Adams. Father Davion had tried unsuccessfully to convert the Natchez.

In 1703 Jean Penicaut wrote in his journal of thirty Natchez villages mostly along the St. Catherine Creek. He also said they were "one of the most polite and affable nations on the Mississippi." He learned later that kindness pays.

De Soto discovers the Mississippi. Courtesy Mississippi Dept. of Archives & History.

FIGHT FOR CONTROL. Meanwhile, France and Spain maneuvered for position. Bienville needed families to occupy the land he was claiming. Some Canadians tried to settle in Natchez but were Indianized as they did not want to stop moving. Bienville struggled with Mobile and finally made some progress before it was moved in 1711. Then the crown grew tired of bothering and leased Louisiana to Anthony Crozat for 15 years in 1713. He chose Antoine de la Mothe Cadillac as Governor and demoted Bienville to Lt. Governor. Bienville became sore and made trouble every chance he could. He nearly caused a mutiny among the troops.

FRENCH BOURBON FLEUR-DE-LYS

FORT ROSALIE BUILT, FIRST OFFICIAL SETTLEMENT. In 1715 Cadillac, returning from his northwest search for silver mines, stopped in Natchez to inquire about the murder of four canadian travellers and an incident in which two brothers, who tried to establish a settlement, were nearly massacred. Because of the incidents Cadillac refused to smoke the peace pipe, then returned to Mobile and ordered Bienville to take fifty men and settle affairs with the Natchez.

Cadillac's daughter had been in love with Bienville, so Cadillac thought it best to offer her hand in marriage. Bienville refused to marry her. Therefore, it isn't too much to assume that Cadillac cared little for Bienville's welfare. But, even with only fifty men, Bienville came to Natchez in 1716, lured the Great Sun and two of his brothers into the French camp and held them as hostages until the Indians put the murderers to death. Finally, Bienville negotiated a settlement which provided for Indian labor to build Fort Rosalie in August of 1716, and he established peace once more. Cadillac was removed, Crozat lost interest and Bienville was Governor again.

THE MISSISSIPPI BUBBLE. France was nearly bankrupt, so a crafty Scotsman named John Law easily convinced the French government to support his scheme of creating the Mississippi Company. The idea was to circulate more money into the economy by issuing paper money guaranteed by the anticipated profits from the vast colonial empire. Profits would therefore pay off the public debt.

In 1717 the Mississippi Company was to operate the government of Louisiana to bring in settlers. Stock was sold and people lined up to pay as much as $25 an acre! Estates were sold in France to obtain colonial land. To become an overnight millionaire was to be called a "Mississippian."

It might have worked, had the bank not had enemies who questioned their ability to guarantee their position. But faith in the bank failed and the people demanded gold for their money. By 1720 the Mississippi bubble burst but France had succeeded in bringing in many settlers.

BLACK CODE. In 1719 a hurricane helped to fulfill Bienville's desire to move the capital from Mobile to Biloxi, and in 1722 to New Orleans. That year the first Negro Slaves had entered the colony. Also that year, France and

Spain were at war again. They traded off Pensacola several times. Meanwhile, the British made firm allies of the Chickasaws and Bienville predicted Indian trouble again. In 1723 Louisiana was officially separated from Canada as a province. Then, in 1724 Bienville gave the colony its first legal code, including the "Black Code" that established conditions for the treatment of slaves.

Fort Rosalie, overlooking Natchez Under The Hill and the Mississippi River. Courtesy Mississippi Dept. of Archives & History.

LAND GRANTED. Louisiana still needed settlers, so the Mississippi Company offered tremendous land grants to get them. Pamphlets and newspapers advertising this were sent out all over France. John Law founded a settlement in the Red River area, purchasing 12,000 subjects. 1600 of them made it there, but when Law was disgraced with the whole mess, the German colonists moved to the Mississippi area just above New Orleans. It is now called the "German Coast." While the settlers moved in on Pascagoula Bay, Bay St. Louis, and the Yazoo Valley, Natchez grew considerably and was considered for the capital.

MASSACRE. Natchez grew so fast that the Indians became very unsettled and took it out on the white man infringing on their lands. In 1723 Bienville had to seek vengeance for a series of murders and ended up killing one of the Suns and some other Natchez. The Indians vowed to return the compliment and did six years later.

In 1724 Bienville was formally removed and called back to France. But by 1729 the Indian nations of the Chickasaw, the Yazoo, the Natchez, and a few Choctaws called a council of war to push out the white man. They set a date for their massacre and each nation was given a bundle of reeds from which a single reed was taken every day to count down to that bloody day.

It is said that "Stung Arm," the wife of a Sun, snuck into the temple and removed some of the reeds. Or it might have been the arrival of a shipment of arms to the fort, that made the Natchez impatient. Consequently, their impatience brought the massacre on sooner than planned.

Chopart, commandant of Fort Rosalie, ignored rumors of Indian trouble and threatened with punishment anyone taking the threats seriously. It was that same morning at 9:00 A.M. on November 28, 1729, that the Great Sun came into the fort, on the pretense of getting ammunition for a hunting expedition, and signaled his warriors for the attack. Nearly every man was killed along with 200 settlers. Only two soldiers survived because they were out in the woods. 250 women and children were captured and only two men who were in the fort at the time were spared for their needed skills: a tailor and a carpenter named Jean Penicaut.

The other Indian tribes attacked as scheduled and there was wholesale bloodshed throughout the countryside. Fort St. Peter in the Yazoo Valley was left in a horrible state of violent remains.

THE NATCHEZ DISAPPEAR. New Orleans heard of the massacre at Natchez and sent 1400 French, allied by 1200 Choctaws, to lay siege on the Grand Village of the Natchez. Negotiations were made on November 25th of 1730, but the Natchez fled across the river the next day. They set up a fort at Sicily Island which was captured by the French in 1732. Only a few escaped to live with the Chickasaws while the rest were sold as slaves in New Orleans to Santo Domingo.

The Natchez were wiped out along with the Mississippi Company, which was funding the lengthy battle.

If you stand on the top of one of the Indian mounds and try to imagine what their nation looked like and how they lived, you might feel sorry that civilized man ever came to this beautiful, natural land and eventually wiped them out. You also might gaze out into the distance or up at the blue sky and wonder if its builders ever made it to the "Happy Hunting Ground." If you believe as I do, you know they did.

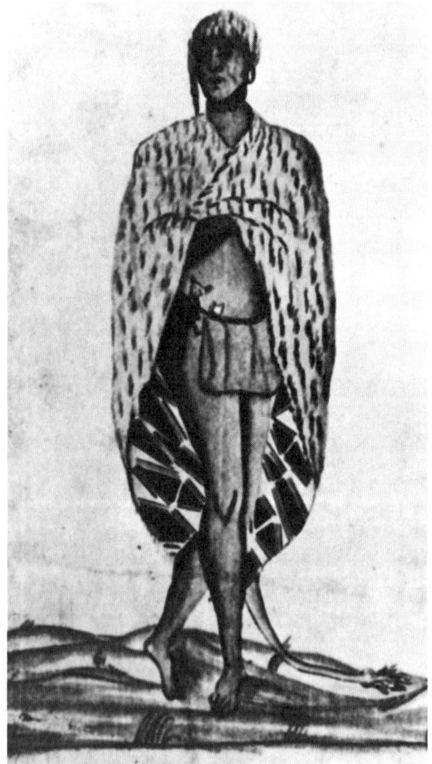

The Natchez Indian–Extinct as a tribe. Courtesy Mississippi Dept. of Archives & History.

FRENCH LAST STAND

In 1732 France took control of the colony and the colonists wanted Bienville back to make settlement with the Indians again. He knew the English had a hand in causing Indian problems with the French, but the Chickasaws, with whom the escaped Natchez found refuge, were the focal point. Beinvielle demanded the Natchez refugees and declared war when the demands weren't met. Drawn out, the war was not settled until 1740 when Binville negotiated peace.

While England had been handing out huge land grants, France sought ways to make the colonies pay off. Tobacco was a good money crop. Cotton was introduced in 1750; the Jesuits initiated sugar cane. Nothing worked and the Indians kept making trouble with English help. Finally, in 1756, the "French and Indian" or Seven Years' War broke out between England and the French-Spanish allies. By 1763 England had gained control of nearly everything east of the Mississippi. Georgia was founded and consisted of land from coast to coast even though some of that land was not owned by England. And in order to appease the Indians, the area westard of the Mississippi was reserved in what the Georgains called "Indian Giving". The northern boundry of West Florida was moved up to from the 31st parallel to the mouth of the Yazoo River, leaving Natchez to its settlers. If not, the land would have been taken away from them and given to the Indians.

JERSEY SETTLERS. Hard times in the 13 Colonies along with the Tobacco Panic of 1768 pushed many settlers westward and into Natchez. By 1770 Natchez was filling fast. England was granting quantities of land to verterans of the French and Indain War according to rank. An officer could recieve 5,000 acres and a private only 50. In 1772, a captain named Amos Ogden received 25,000 acres in Kingston on the Homochitto River just outside of Nacthez and, with two wealthy New Jersey planters named Richard and Samuel Swayze, started the "Jersey Settlement". Anthony Hutchins settled Coles Creek in 1773. When the American Revolution broke out, loyalists to the crown were given refuge in Nacthez. Sir William Dunbar, a prominent Scotsman from Pennsylvania was one of them. By 1776 the town of Natchez was etablished.

RATTLETRAP. Another man from Pennsylvania who started a store at Natchez Under The Hill in 1774 was James Willing, who lived extravagantly and became indebted. In 1778 Willing, with secret support from the Continental Congress, led thirty men and an armed ship called "Rattletrap" from Fort Pitt (Pittsburg) to pick up Sapnish supplies and secure neutrality of the people of West Florida. Willing had complied a "Black List" of British Loyalists which was his reference of action. When he arrived at Hutchins' settlement at Coles Creek, Willing captured Hutchins and took possesion of his things. Others in the area fled or had to take a loyalty oath. When Willing made it to New Orleans he sold his booty.

The English Governor of Florida, Peter Chester, was angered at the Americans and Spanish so he sent two war ships to New Orleans and troops along the Mississippi. The Loyalists in Natchez revolted and Willing was eventually captured. There was confusion and conspiracy in Natchez until the British took over in 1798.

SPANISH CASTILE & LEON

SPAIN STEPS IN. Meanwhile, Spain declared war on the British in 1779 and Bernardo de Galvez, the Spanish Governor of Louisisna, Captures Natchez and the Mississippi by September 21. He laid siege on Mobile and on Pensacola. The Loyalists took his preoccupation with Pensacola as an opportunity to revolt and regain control. While the confusion was hieghtening in the area, Georgia took the opportunity to recalim their orignial stake in Natchez and formed the Bourbon County around Natchez and tried to dispose of the land. The commissioners from Georgia were run out of town.

Uncpposed, Spain moved back in 1781. The city of Natchez was nearly deserted as Loyalists refugeed to places like Savanah and some even to join the Chickasaw Indians and a man named James Logan Colbert. Colbert pirated Spanish ships so well that Spain invited them to come back to Natchez.

MANUEL DE GAYOSO. The Spanish tried very hard to please the people in their territory in order to avoid insurrection. Natchezians liked their Spanish Commandant, Manuel de Gayoso, who had an African wife and an English education. He was pleasant and gave many lavish parties. He built the magnificent home "Concord" and named it for the peaceful existence he wished to establish. It was Gayoso who had the Spanish engineer, Collel, lay out downtown Natchez in square blocks. By 1797 Gayoso was Governor of Louisiana.

BIT. Spain conducted trade in most of the Lousiana territory even though English merchants set up floating stores which sold goods from boats to avoid Spanish restrictions. Still, there was more Spanish money in circulation than British. The term "bit" came from the division of the Spanish dollar. One "bit" was one piece of the "pieces of eight".

THE WHITNEY GIN. Spain still sought the money crop which would make the colonies profitable. 1,500-2,000 pounds of tobacco could be grown on one acre around Natchez but it was considered inferior in quality compared to that produce in Virgina and Kentucky. The Spanish warehouses overflowed and General James Wilkinson issued duty-free permits in Kentucky which caused the Natchez planters to turn to another crop- Indigo. A valuable dye growing at 150 pounds per acre, but expensive and complicated to process, Indigo also was hard on the health of the slaves, polluted the streams, and was easily consumed by insects in 1793. Planters were relieved that same year by Eli Whitney's invention of the efficient cotton gin.

In 1795 Daniel Clark and two mechanics reproduced the Whitney Gin from a traveller's description and a crude drawing. It could gin 500-1,000 pounds per day. Immediately, the Gins were manufactured and sold in Natchez. Cotton became the "universal crop of the country" according to Sir William Dunbar who built and lived in a home he called "The Forest." Cotton was so important that gin reciepts were accepted as paper money because there was very little money in circulation in this area. Cotton was packed in bags and then wooden boxes until Dunbar devised a square bale around 1799.

CHRIST CHURCH. Spain was fair in most respects but they did not allow public Protestant worship. At one time there were rumors that all the Protestant

Bibles would be burned. Samuel Swayze hid his Bible in the hollow of a sycamore tree. In 1791, Reverand Richard Curtis Jr. organized the Salem Church, the first Baptist church in Mississippi, but when a Spaniard was converted to Baptist in 1793, the Government ordered his work to stop or be expelled. He left.

Episcopal clergyman Adam Cloud, who was the first Epicopalian minister in Mississippi, was sent in chains to New Orleans. Yet, he reappeared in 1820 to organize Christ Church at Church Hill near the Trace.

The Spanish also tried to organize schools. Gayoso said they would be useful in teaching the younger generation to love Spain. That too failed.

SPAIN LOSES FOOTING. In the Treaty of San Lorenzo of 1795, Spain agreed to the boundry line of 31° latitude. It made Natchez and everything north American. It opened up the river and the port of New Orleans to Americans. However, the Spanish lingered in hopes that the American government would collapse.

Andrew Ellicot, an astornomer and American Commissioner, arrived on February 24, 1797 tosurvey the 31st parallel. Sir William Dunbar, the surveyor employed by the Spanish to carry out a provision of joint survey, assisted Ellicot

Antebellum "Christ Church" at Church Hill.

at Connolleys Tavern. At the protest of Gayoso, Ellicot raised the American Flag over his quarters.

Gayoso and wife Margaret Watts very much wanted to live in the home they were constructing, Villa Gayoso, but soon had to leave. The "Council of Gentlemen", attended by pro-American revolutionaries such as Anthony Hutchins, "Squeaking Tony, long and bony", and Thomas Green, spoke out against the Spanish remaining.

In April of 1797, Piercy S. (Crazy) Pope, arrived with a small detachment of men and set up camp outside Fort Rosalie with assertions of taking and occupying it. His American force borowed tents from Gayoso to camp. Ellicot, who also overstepped his authority, argued with Pope on who was in charge of things and Gayoso aggravated the situation with rumors of Indian uprisings and British invasion from Canada. To relieve the tension Gayoso issued a permit to John Hannah on April 9 allowing the Baptist minister to hold services, except for mentioning any politics. Hannah forgot his promise and critisized the Spaniards and their religion. Some Catholic Irishmen in Natchez Under The Hill rioted. Hannah demanded justice but Gayoso arrested him. As Hannah was dragged past the American encampnemt he yelled, "Help me, citizens of the United States!"

STARS & STRIPES

Tension mounted and the Americans threatened to attack. Gayoso set up for the defense but went off to become Governor of Louisiana. Assuming control of the city was the schoolmate of Andrew Ellicot, Stephen Minor. Then Captain Issac Guion came with orders to take over the fort. He issued an order that the Spanish be out by March 31 and on March 30, 1798 the Spanish left never to return.

MISSISSIPPI TERRITORY. The next week Congress set up the Mississippi Territory. James Wilkinson assumed control at Fort Adams. The first Governor of the territory was Winthrop Sargent, who the country people called "His Yankeeship". He set up Adams county. When Jefferson replaced him with W.C.C. Claiborne, Sargent moved to Natchez and built the beautiful home called "Gloucester." In 1802 the little town north of Natchez called Wshington was chosen as the capitol, which was supposedly far enough away to be safe from the influence of the wealthy classes of the former capitol.

THE NATCHEZ TRACE. During the Claiborne period, Natchez was suddenly opened up to the world of trade. It was in 1801 that a treaty was signed with the Chickasaws and the Choctaws to open the Natchez Trace, an overland trail which led from Natchez to Nashville. At some places, pack horses had to be used. A trip from Nashville to Natchez took ten days and four hours. It was so slow and cumbersome that a man named Walking Johnson beat the Trace Postal Service on three seperate occasions.

There was also the Gaines Trace, which ran from Muscle Shoals, Alabama to Cotton Gin Port and later to the Natchez Trace at Houlka. The Three-Chopped

Way (because it was marked in 1807 with three notches) ran from Natchez to Milledgeville, Georgia. Here Sam Dale became famous as a scout, a guide and operated a wagon train. After the War of 1812 the Government sent Andrew Jackson to build a military route from Columbia, Tennessee to Madisonville, Louisiana.

On these lonesome routes, villains fond their trade easy. The Natchez Trace in particular became infested with evil. A prticularly notorious bunch consisting of a man and his sons Samuel Masons Gang, who were recognized and whipped on the pillary in Natchez. Thereafter, they scrawled the name "Mason" in their vitims' blood.

Other criminals also appeared on the trace like the "great western land pirate," John Murrell and his "mystic clan" of nearly 500 slave stealers who rarely missed any opportunity for crime. Many of them were hanged in the 1830's.

LOUISIANA PURCHASE. Spain withdrew the Americans' depository rights in New Orleans and Mobile was also closed so that the Mississipi territory was "landlocked". President Jeffferson did not allow this and sought to purchase Louisiana. Since Spain was running Louisiana for France and Napoleon needed money for a fleet of ships, the purchase was made in 1803. New Orleans and the entire Mississippi Valley were open for western trade. Claiborne was moved to Louisiana Governor and Cato West, territorial secretary, took over. In 1805, Robert Wiliams relieved him. When the War of 1812 broke out, Mississippi's "Most Able Territorial Governor", David Holmes was in office. With the end of the war came the end of all the Indian troubles and Mississippi grew large enough to be a State.

River Front at Natchez Under The Hill. Courtesy Mississippi Dept. of Archives & History.

DEVELOPMENTS

COTTON IMPROVED. Cotton prices grew as high as 15 cents a pound. Sir William Dunbar's son-in-law, Samuel Postlewaite, invented a harrow to plow cotton stalks from the ground instead of burning them. And he also solved the problem of disposing cottonseed by using it as feed for cattle and fertilizer for corn. They were later able to use the cottonseed oil and opened mills below the bluffs.

Also, in the beginning, most of the cottonseed tried here were no good. The upland breeds rotted in the humid Natchez climate. Others were not as good as the Mexican seed which Dr. Rush Nutt improved to what is known as the "Petit Gulf" variety. Story has it that Mexico did not want their cotton grown anywhere else and General Wilikinson, sent Walter Burling to Mexico to collect a $121,000 bill for services to the King, and smuggled out some of the Mexican seed in toy Dolls. Dr. Nutt supposedly used the seed to hybrid his new variety.

Taking Coal at Natchez. Courtesy Mississippi Dept. of Archives & History.

SLAVERY OUTLAWED. Labor was needed to pick cotton and some Choctaw Indian women were used until the Constitution ended American slavery by 1808. There was some slave smuggling from the Spanish territory, but it was easier to get slaves from the upper south which had more than it needed.

STEAMBOATS. Flatboatmen brought goods from the north to sell to the needy south. At one time there were 83 of them at Natchez Under The Hill. They would float down stream to New Orleans, sell everything (including their "Kentucky Ark"), ferry back to Natchez and walk the Trace back up north, only to be robbed, then start the journey over. Trying to navigate upstream was to use poles to push, ropes to pull and even grab branches on the riverbank and "bushwhack" against the seven mph current.

In 1811 Nicholas Roosevelt built the first successful steamer to navigate the Mississippi–"The New Orleans." It made the trip from New Orleans to Natchez pleasant. Then Henry Shreve, the "Father of Mississippi Steamboating," perfected his broad vessel with a shallow draft which could navigate the worst snags in the river making river trade boom.

PRINTING Mississippi's first printing press was brought in by Andrew Marschalk in 1798 when he printed a ballad. In 1799 he moved to Natchez to publish Sargent's first laws. He published Mississippi's first paper, "The Mississippi Gazette" and later the "Mississippi Herald." He also published a book by Scotch Presbyterian, John Henderson–"Paine Detected."

EDUCATION & SCIENCE The first school for girls was established in Natchez in 1801 by David Ker, a Scotch-Irishman who founded and served as president of the University of North Carolina. In 1811 Jefferson College was opened in what was James Smylie's Washington Academy. The Methodists set up one of the first Sunday Schools in the south at Natchez in 1826.

In 1803 a group of gentlemen chartered the Mississippi Society for the Acquirement and Dissemination of Useful Knowledge. Sir William Dunbar, one of the founders, has been noted for many scientific accomplishments including more than a dozen papers on Indians, fossils, eclipses, and comets. He also explored the Red and Ouachita Rivers, including the Hot Springs area in Arkansas. There was also Solomon Turnipseed, who humorously but accurately blamed the mosquito for Yellow Fever.

Cotton loaded on Paddle Wheeler at Natchez Under The Hill. From the Prince Collection.

POLITICS. In March of 1817, after the President Madison signed Mississippi's right to become a state, a convention was held at the Methodist Church in Washington. New boundaries to the state had already been drawn. Mississippi had lost part of their backwoods to Alabama because the people felt the wealthy Natchez area maintained too much voting power over their remote settlements. Natchez was finally made the state capital because the Natchez element controlled the convention. George Poindexter chaired a 21 man committee to draft the state constitution and the legislative, executive, and judicial branches were formed. In September of 1817 David Holmes was elected the first state Governor.

Desire for a more central capital drove the legislature to seek out and finally choose the Jackson area, named after Andrew Jackson. The first legislature met at Jackson in a frame house in December of 1822.

Democrats and Whigs were the political parties which were adapted and Natchez was a Whig stronghold.

SLAVERY RETURNS. In the 1820's slave trade into Mississippi got out of hand until the 1832 constitution prohibited the introduction of slaves into the state as merchandise for sale. There was an occasional free Negro. They had to carry Free Papers and be of good character. William Johnson, a free black man who was a barber in Port Gibson, came to Natchez in 1830 and acquired three barber shops, a toy store, a drayage business, and even a small loan service.

In the 1830's good and bad luck came at once. Cotton exploded and began to push Natchez into the view of the world while the activities of Northern Abolitionists provoked slave insurrection. In 1836 there was a crisis involving several slaves who were hanged along with the white men who stirred them up. To make matters worse, while planters were only beginning to see the potential of the cotton kingdom, the nations economic stability collapsed in the panic of 1837. The good growing season here could not be realized. To top it off, there was a tornado in 1840 which reigned destruction, including the annihilation of Natchez's only light house.

Then in 1846 the legislature reopened slave trade because of a lack of the needed manpower.

BANK OF THE MISSISSIPPI. Back in 1809 the Bank Of The Mississippi had been set up in Natchez. In 1818 the institution was made a partially state-owned bank and granted a monopoly, but in 1821 Governor George

"The Father Of Secession"
Governor of Mexico City and Mississippi, General John Anthony Quitman.
Courtesy Mississippi Dept. of Archives & History.

Poindexter suddenly decided that the monopoly was unconstitutional. Poindexter killed one of the bank directors in a duel but some said he fired too soon. He was not supported in Natchez.

NEW THINKING

Duelling was forbidden by the 1832 constitution and allowed for heavy fines for duelers. Duelers were forbidden from holding public office in the state. Also, the survivor of a duel was required to pay off all of the debts of the person he had killed.

THE FATHER OF SECESSION. In 1821 John Anthony Quitman came to Natchez from New York to set up a law practice and live in a mansion called "The Monmouth." Known as the "Father of Secession," Quitman was elected into the legislature in 1827, where he proved to be a strong state-rights leader. In 1835 he became the leader of a society to suppress gambling and other vices. Many professional gamblers were driven from the city. In Vicksburg a similar society had to use bloodshed to expel a group of gamblers who were then chased out of town. Also to clear the soul in 1837, The Cold Water Man, a temperance journal, was founded in Natchez.

In 1836 General Quitman led a group of volunteers to assist Texas in their struggle for independence. He was the leader of the troops who took Mexico City and won the war against Mexico. He was elected Governor in 1849. Quitman fought hard for secession even to go so far as to support an expedition to Cuba led by Narciso Lopez, a leader of Spanish refugees in the U.S. Stating that Cuba was ready to revolt against Spanish oppression, many southern volunteers went with him, hoping to bring Cuba under U.S. control as a slave state. Quitman was indicted and then arrested in 1851 and resigned his office. Charges were dismissed in March of 1852 and rejoicing was heard in Mississippi. Natchez fired a 15-gun salute for Quitman and another for the

"Monmouth," home of General John A. Quitman.

15 southern states.

Former Whigs who formed the Unionist Democrats and favored staying in the union, nominated Senator Henry Stuart Foote as their candidate for Governor. Quitman, who had quickly returned to politics, opposed and was supported by Jefferson Davis and the "Fire Eater" Democratic faction. The candidates toured the state peacefully until Foote began a mudslinging campaign hinting that his life was threatened and he expected to be assassinated before the end of the campaign. He mentioned Quitman's Cuban interests. Quitman denounced the accusations as false and cowardly and ended the joint tour. Finally, a fight broke out between the two.

Election for secession came and the Unionists won by 7,000 votes. Quitman was furious and withdrew from the Governor's race. Jefferson took his place even though he was sick and unable to campaign vigorously. Foote won the race by 999 votes. Succession came only after Quitman had died.

ANTEBELLUM NATCHEZ. To color the attitude of the day, here is the Antebellum Creed, as the lawyer Reuben Davis wrote in his Recollections of Mississippi and Mississippians: "*There creed was generally simple. A man ought to fear God, and mind his own business. He should be respectful and courteous to all women; he should love his friends and hate his enemies. He should eat when he was hungry, drink when he was thirsty, dance when he was merry, vote for the candidate he liked best, and knock down any man who questioned his rights to these privileges.*"

Elegant home "Arlington" indicating Antebellum prosperity.

Antebellum Natchez was the cultural center of the south. There was the theater, great artists such as John J. Audubon, elegant architecture, entertainment, folklore, circuses and showboats. There were also many writers. A lawyer named Joseph Holt Ingraham came from New England in the 1830's to teach at Jefferson College also published his first book of short stories called "The Southwest, By A Yankee." He wrote dime novels in serial form which later came out in paperback, like the one about Jean Lafitte, the pirate. In only a few years he wrote 80 novels. One year he wrote 20.

The Hustle and bustle of old time Natchez as cotton goes out and people come in. From the Prince Collection & Natchez Photo Lab.

Ingraham moved to Nashville in 1849 and became a clergyman. In 1851 he founded a church in Okolona, Mississippi and served at Aberdeen and Holly Springs, where he also operated a boys' school. Ingraham later wrote religious novels like "Prince of the House of David." His son, Prentiss, wrote 600 novels and 400 novelettes.

Other scientific work besides that of Sir William Dunbar came from the area. Dr. John Wesley Monette, of Washington contributed to medical periodicals and wrote a six-volume "Physical Geography and Geology of the Mississippi

Valley in 1846. He also wrote many articles about yellow fever and, in 1841, had the city of Natchez adopt a quarantine to prevent its spread. Other doctors at the time laughed at him for such precautions.

STARS & BARS

CIVIL WAR. When the Civil War broke out in 1861 Natchez had to forget its culture for a while, but it was not lost. Even on a battlefield somewhere in Virginia, a young Captain T. Otis Baker from Natchez read books and pondered when he could. In his diary in 1864 was a rule for self-examination: "*What new ideas–What new proposition of truth, have I gained–What further confirmation of known truths–and what advances in any part of knowledge have I made?" There must have been many more like Baker, and they were ready to help Mississippi win the peace, even if they could not win the war.*

The Natchez Rifle Group. Courtesy of Bill Stewart.

The war only briefly visited Natchez. There was no military defense but Grant's army set up a command post in the magnificent home "Rosalie," which stands very near the exact location of the original Fort Rosalie. There were only a few incidents as when the gunboat Essex fired upon the city, and one of the Natchez mansions was blown up to make room for a Federal fort.

OLD GLORY

FREEMEN. The slaves were set free but most of them stayed here and even remained loyal to their former masters. The Confederate soldiers, torn and tattered, dragged home to a city under a new flag.

There were many stories about the cruelty of the Union Army that went through Georgia, but the people of Natchez only knew of a brave troop of men, some of them under the command of General Ulysses S. Grant, who were strong in force and belief and very big in mercy. Because the beaten Confederate soldiers were able to come home to the beautiful city and homes they had left and allowed to live in peace, they were able to finally realize the truth of it all.

They fought for a cause, not to own another human being, but not to be told how they had to live their lives. This was the cause that got distorted into something compelled by greed for money and power. This was the cause that was perverted, by the words of fear, into a hatred of men by themselves. This was the cause that sent men to fight men, households to be divided, brother against brother, sister against mother and father against son. This was not the cause that was lost, but that was found and clarified through bloody battle so that all could see and understand it in its absolute truth.

"We hold these truths to be self-evident, that all men are created equal; that they are endowed by their Creator with certain unalienable rights; that among these are life, liberty, and the pursuit of happiness...

"The Declaration of Independence

"We, the People of the United States, in order to form a more perfect Union, establish justice, insure domestic tranquillity, provide for the common defense, promote the general welfare, and secure the blessings of liberty to ourselves and our posterity, do ordain and establish this Constitution for the United States of America."

The cause was: not to be told how to live or to worship God or what to read, nor to be owned by other men. The cause was Liberty–the Freedom of all people.

Grand Army of the Republic. Courtesy of Bill Stewart.

Unfinished Antebellum Home "Longwood" Photo by Mark Coffee.

2

ANTEBELLUM NATCHEZ

The glory of Natchez lies between the Old Natchez and the New Natchez. It is part of the past and the present at the same time. It was here before any of us were born, and it appears that it will go on forever. This aspect of Natchez is a part of the living past. That living past is the homes, streets, buildings and stories we live in and around, which came to us as the gesture of a culture, the presence of a people that once were, people who left their mark on the face of the earth before they passed through to another time. Here they left the evidence of their lives and the way they lived as a testimony to living life gracefully. That memory lives on in splendor.

Antebellum Home "Arlington."

Considerately, they picked a spot and planted their testimony here so that all those passing through time could see and have a reference point to the culture they wished to preserve. That graceful gesture to their children and descendants allows us to stand here today and see them very clearly in their own lives in the past, and feel the timelessness that is the human spirit. Take a step back in time with us and venture into those places that exist today and in the past.

There are over 500 Pre-Civil War buildings in Natchez, and the dates are often vague, so this alphabetical list is by no means complete or perfectly accurate. It is enough, however, to gain an idea of the histories of the homes and may help you to choose which homes to tour. The downtown houses are easily found by referring their numbers to the map in Chapter 5. Bed & Breakfast and Tour information is found in chapter 4.

"Auburn" in Duncan Park.

1. **AGRICULTURAL BANK** (1833), 422 Main St. Its erection introduced Natchez to the Greek Revival Style with this supreme example of architecture. Now it's the Britton & Koontz First National Bank.
2. **AIRLIE** (1790), Elm St. One of the oldest homes in the area and built before the date shown, this charming cottage was used by the Union soldiers as a hospital.
3. **ARLINGTON** (1820) John A. Quitman Pkwy. Exquisitely done by Jane Surget White, the wife of Architect Captain James Hampton White, the magnificent palace has 17-foot ceilings and an almost endless collection of heirlooms.
4. **ARMSTRONG HOUSE** (1838), 613 Washington St. This house was built by famous local builder/developer E.P. Fourniquet.
5. **AUBURN** (1812), 400 Duncan Ave. Once owned by Dr. Stephen Duncan, one of Natchez' first millionaires, who had to leave during the war because of his abolitionist sympathies, this National Historic

Landmark was deeded in 1911 by his family to the city for a recreation park. The house was constructed by Levi Weeks for Lyman G. Harding, the first Attorney General of the Mississippi Territory. Charming its grounds is a 350 year old live oak tree registered with the American Live Oak Society. Gracing the entrance hall is a spiral, geometrical stairway which stands unsupported.

Operated by the Auburn Garden Club, 442-5981. Open Mon.-Sat. 9:30 AM-5:00 PM and Sun. 1:30 PM-5:00 PM. Admission: $4.00, children age 6-18, $2.00. Year round Bed & Breakfast, seated dinners, luncheons, and Champagne Tours by reservation only.

6. **AUNT HILL** (Circa 1828), 610 Jefferson St. This frame house appears today as it did so long ago when it was built.

7. **BANKER'S HOUSE** (1838), 107 S. Canal St. Built to house the banker of the First Bank Of Commerce, with the master bedroom directly on top of the bank's vault, the house was once a girl's school and a boarding house. As evidenced by the register, Frank and Jesse James slept in this National Historic Landmark in 1887. The house has a slate front walk that was used as ships ballast. Also contained within is a square rosewood grand piano made in Copenhagen in 1843, Claiborne's Mississippi History, all 13 books of Audubon's Indians, Birds and Quadrupeds, and a wonderful primitive painting of pre 1840 Natchez on a linen bed sheet by Rebecca Gustine Minor.

8. **BARNES HOUSE** (1830-1840), 705 Washington St. Built by E.P. Fourniquet, this lovely home has a beautiful side garden.

9. **BELMONT** (1840), Hwy 61 S. Built very strong and sturdy after the tornado of 1840 by Loxley Thistle, this lovely home was once named Clermont.

"The Briars" Courtesy of The Natchez Convention Center.

10. **BELVIDERE** (1838), 70 Homochitto, was the house of Gayoso's secretary and contains a 1753 silhouette of the mayor and his wife.

11. **BIGGS HOUSE** (1796), 300 N. Pearl St. This Greek Revival house was once a carpenter shop.

12. **BRANDON HALL** (1856), Natchez Trace. Out of the 28,000 acres which made up this estate, Gerard Brandon carved out a fortune and left his son, Gerard, with the task of becoming the largest indigo and cotton planter in the south and the first native-born Governor of Mississippi.

"Cottage Gardens"

13. **THE BRIARS** (1818), behind Ramada Hilltop, was where Confederate President Jefferson Davis married Varina Howell, "The Rose Of Mississippi".

14. **BRICK TOWN HOUSE** (1845), 312 N. Union. This two-story building is the last remaining section of the three connected segments that were originally built.

15. **THE BURN** (1836), 712 N. Union St., is called so after a stream that ran through the estate. The Scotch meaning "The Brook", was ironic when the top floor really burned and left the changed home the way it is today. The house was used as a Union hospital and headquarters. There is a photograph of General U.S. Grant at this house which has belonged to three Natchez' Mayors including Tony Byrne and his wife, who now occupy this elegant home.

Open daily 9:30 AM-5:00 PM, there is a swimming pool and beautiful gardens for the Bed & Breakfast guests, dinner parties, luncheons and picnics may be arrange for groups of 20 or more. (800) 654-8859, 442-1344.

16. **THE CEDARS** (1814), Hwy 553, Church Hill. The grounds of this Greek Revival structure are open for drive through.

17. CHEROKEE (Before 1836), 217 High St. Built under the Spanish ordinance for Ebenezer Rees, this house has its main building on top of the hill with its gardens and back rooms further down the slope. It was believed that the dreaded yellow fever was released from the ground and no excavation was allowed.

The house was later owned and occupied by the Frederick Stanton family while waiting for Stanton Hall to be completed. It is now owned by the Junkin family who's ancestors lived at Sunnyside and Homewood.

18. CHERRY GROVE (1788/1860's), Hwy. 61 S. Pierre and Catherine Surget came to Natchez on a sailing ship and traded their cargo of pig iron to a band of Indians for a piece of land at Second Creek, where their first home was built. When it was destroyed by fire in 1860's, they rebuilt.

19. CHOCTAW (1836), 310 N. Wall at High St., was the first free school in Natchez and was built by Joseph Neibert. It served as Stanton College for a time. Now it is owned and operated by the Natchez Art Association.

20. THE CLIFFS (1850), Hwy. 61 S. Beautiful Greek Revival Plantation house.

21. CONCORD (1788) Though it was destroyed by fire many years ago, this magnificent mansion deserves mention because it was the lifetime dream of Don Gayoso de Lemos, the Spanish commandant, and named for the peace which he hoped it would project.

"Edgewood" Courtesy of the Natchez Convention Center.

22. **CONNELLY'S TAVERN** (1798), 200 N. Canal St. Is called The house on Ellicot's Hill as it was built on the spot of the surveyor's encampment on the 31st parallel. It was once the home of Dr. Frederick Seip, founder of the Natchez Hospital and was the Natchez High school before the Civil War. A National Historic Landmark, this unique structure sports a dry moat with draw bridges for the protection of its guests. (800) 647-6742, 442-7259, 442-6672.

23. **COTTAGE GARDENS** (Before 1836), 716 Myrtle, was built of ships timbers on the property of Don Jose Vidal.

24. **COUNTY COURT HOUSE** (1820), Courthouse Square, Federal Style building was remodeled around 1920 in Neoclassical Revival.

25. **COYLE HOME** (Circa 1800), 307 Wall. Possibly much older than the date indicated, this frame house might have had its basement added when the street was leveled in the early 1800's.

26. **D'EVEREUX** (Circa 1836), D'Evereux Dr. Site of the filming of "The Heart Of Maryland", this magnificent palace is one of the most awe inspiring homes, in a perfect setting. It was built by William St. John Elliot and visited by many famous people, including Henry Clay and also the Union Army, who camped on her grounds, supposedly right on top of the buried valuables. It was here that the two murderers of George Washington Sargent of Gloucester were hanged.

    D'Evereux is the Elliot family name. A relative, Gen. John D'Evereux, fought with Simon Bolivar in South America.

27. **DIXIE** (1853), 211 S. Wall St., the back building was built by Maurice Stackpoole, but the front was added later by Samuel Davis, the brother of Jefferson Davis. The fine cottage was also owned by the Bowies, family of Jim Bowie.

28. **DR. DUB'S TOWNHOUSE** (1852), 311 N. Pearl St., was the home of City Alderman and Dentist, Dr. Charles H. Dubs, who invented the Compound Union Screw Forceps to extract hollow roots safely.

29. **DUNLEITH** (1856), 84 Homochitto. It was Charles Dahlgren (who's father was the first U.S. Consul from Sweden and descendent of the King of Sweden, Gustavus Adolphus) who married Mary Routh. As a wedding present, the millionaire Job Routh, gave them the land on which Routhland was built. When lightning burned it to the ground, they built this magnificent Greek Revival manor with its 26 Tuscan columns. Alfred V. Davis, raiser of thoroughbreds, bought and renamed the home. When Union soldier came to acquire horses, Davis supposedly hid some of his prized stock in the basement directly beneath the dinning room, where the soldiers ate. It was later acquired by Joseph Carpenter and remained in his family for four generations.

    It was the Dahlgrens' daughter, Sarah Dorsey who willed her Biloxi home, Beauvoir, to Confederate President Jefferson Davis.

    This National Historic Landmark is available for Bed & Breakfast. (800) 647-6742, 446-6631. AAA 4-Diamond rating.

30. **EDGEWOOD** (1859), Pine Ridge Rd. This unpretentious yet luxurious home lies on part of the extensive John Bisland estate. Given to his daughter, Jane Bisland Lambdin, wife of Samuel Lambdin, the home had a basement kitchen with dumb-waiter to the dinning room and indoor storage tanks for the piping of water to the rooms.

31. **ELGIN** (1792/1840), off 61 S. in the Second Creek neighborhood, was given to his granddaughter, Annis Dunbar, by Sir William Dunbar of the nearby Forest Planation, who received the plantation for his efforts as surveyor of Natchez for the Spanish Government. Dr. John Carmichael Jenkins, who married Annis, was one of the first to study soil depletion, experiment with fertilizers and the packing of fresh fruit in ice for shipment.

32. **THE ELMS** (Circa 1805), 215 S. Pine. Erected during the Spanish influence, this beautiful home with its wide verandas and beautiful gardens sits in a shaded setting, a picture of peace and grace.

Among those who lived in the house were John Henderson, who wrote the first book printed in the Natchez Territory; Rev. George Potts; David Stanton, brother of Frederick Stanton of Stanton Hall; the Drakes and the Carpenters and Lewis Evans, first Sheriff of the Mississippi Territory.

33. **ELMS COURT** (1830's), John R. Junkin Dr. Evidence shows that there was an original house here around 1810 but the Evans family built the 22 room Venetian type structure that still stands unchanged. It contains six gasoliers and a seven octave piano. This beautiful house once belonged to one of the first millionaires of Natchez, Frank Surget, and to his daughter Jane and son-in-law, Ayres Merrill, Minister to Belgium.

34. **ELWARD** (1844), 612 Washington St., still stands as a pure and architecturally perfect beauty since Richard Elward built it.

35. **EVANS-BONTURA-SMITH** (1851), 107 Broadway St. Built before 1790, the back of the house was added on to some time before the Civil War. Inside is the bed Scarlet O'Hara used in "Gone With The Wind".

36. **FAIR OAKS** (1822), Hwy 61 S. was built by Henry W. Huntington after he married Helen Dunbar of nearby Forest Plantation. They called it Green Oak. The pretty home which was called Woodbourne at one time has a 95 foot gallery across the front. Later purchased by the Metcalfe family and renamed Fair Oaks, this pleasant plantation has remained in the Metcalfe-Lanneau family ever since.

37. **FIRST CHURCH OF CHRIST SCIENTIST** (1838), 206 Main St. Once the Commercial Bank connected to the Banker's House on Canal St. This stone-looking stuccoed structure is finished in grey marble, Greek doric portico with ionic columns.

38. **FIRST PRESBYTERIAN CHURCH** (1828), 117 S. Pearl St. Said to be the finest Federal Style church in Mississippi, this architectural pleasure was the model for the church trial scene in Mark Twain's "Tom Sawyer".

39. **FOREST HOUSE** (1788/1810), Pine Ridge Rd. The original hand-hewn log cabin was built before 1788.

Looking north up Pearl St. Magnolia Hall on the right and the Firtst Presbyterian Church steeple in the distance. Courtesy of Bill Stweart.

40. **FORT ROSALIE** (1716), the first known structure built by civilized man was a garrison which was rebuilt many times and was most strategically located on the hill behind the present mansion Rosalie. It commanded views of many miles in both directions of the river, out across the river over Louisiana, and over the countryside that is now downtown Natchez.

To accommodate the French settlers after Fort Rosalie was built, many buildings, stores and houses were constructed, but they were frame houses which could not stand the weight of time. Records and writings tell of a thriving settlement that has now withered away into dust. It wasn't until that dust began to produce crops that permanent structures were erected on the plantations.

41. **THE GARDENS** (1794), Cemetery Rd. is a lovely planter's home with a wide gallery across the front and beautiful gardens.

42. **GLENBURNIE** (1833), John R. Junkin Dr. This Federal style beauty is the site of the famous Goat Castle Murder.

43. **GLENFIELD** (1812), This house was made in a Gothic design of red brick and hand-hew timbers. It was also the scene of a Civil War skirmish as evidenced by the bullet holes in the exterior walls.

44. **GLOUCESTER** (Circa 1835), 201 Lower Woodville Rd. The home of the first territorial Governor of Mississippi, Winthrop Sargent, this home, once called Bellevue, was bought from David Williams, enlarged and renamed by Sargent. A veritable fortress, with iron bars and dry moat, it was also the burial place of Sargent's son, George, who was shot by marauders when he opened the front door.

45. **GOVERNOR HOLMES HOUSE** (1794), 207 S. Wall St., Once owned by Jefferson Davis, this was also the home of the first Governor of the State of Mississippi, David Holmes.

46. **GREEN LEAVES** (1838), 303 S. Rankin. The rebuilders of this home, Jonathan Thompson and his family, all died one week of a yellow fever epidemic in 1820. It now belongs to the Koontz-Beltzhoover family who have contributed and acquired an abundance of family heirlooms, including a set of china that was hand-painted by John J. Audubon and a French sword from the battle of Waterloo given to Col. Daniel Beltzhoover by a dying English soldier at the siege of Vicksburg.

47. **GRIFFITH-MCCOMAS HOUSE** (Circa 1800), 301 S. Wall St., was lifted to add the bottom floor.

48. **THE GUEST HOUSE** (1840), 401 Franklin St. This Colonial Revival Style cottage was redone by the Elks Club, and architect David Peabody built an award-winning glass garden pavilion here when the Eola Hotel began using it as a guest house.

"Dunleith" Courtesy of Natchez Pilgrimage Tours.

49. **HAWTHORNE** (1820's), Lower Woodville Rd. is situated on a land grant in 1786 to Benjamine Bleak, who held the first American Political meeting in the area at his home. George Overaker bought some of the land and built this beautiful plantation home. After Bleak died, Robert Dunbar purchased it, and a descendant of his, who married Douglas S. Bisland, sold it to Leonard Koerber. It went to the Hamiltons and the Hansens and the McGehees. It is now owned by Bettye McGehee Jenkins and her husband, Hyde Dunbar Jenkins, descendant of William Dunbar of The Forest Plantation and Dr. John Charmichael Jenkins of Elgin Plantation.

50. **HOLLY HEDGES** (1796/1832), 214 Washington St. Built by Don Juan Scott with the stipulation that he was not to hold bullfights in the yard.

51. **HOMEWOOD** (1858) Destroyed by fire in January of 1940, yet deserves mention out of its magnificence. The mansion was built as a wedding present by cotton king, David Hunt, for his daughter Catherine and her husband William Balfour. It was built with 100,000 bricks, silver hardware, indoor plumbing, Belgium made ruby glass around the entrance door, Spanish fluted ionic columns, Italian marble, and the top hall was octagonal. It was one of the greatest but is gone like the times it was part of.

52. **HOPE FARM** (1780's), 147 Homochitto, was build by Marcus Hoiler. In 1779, Spain declared war on the British and quietly took control over the Natchez area while the American government was still busy in the east. It was Manuel de Gayoso, the Spanish Commandant, who added the front of Hope Farm in 1789 and laid out the Street Plan of downtown Natchez. And it was Katherine Grafton Miller who restored the home and in 1932, with some of her friends, founded the International Natchez Pilgrimage with the invitation: "Come to Natchez, Where the Old South Still Lives." She traveled throughout the country promoting tourism to Natchez and pleading for the opening of the great homes which grace this little city.

For this and endless other achievements and awards the Mississippi State Legislature adopted a resolution commending her life and named Nov. 12, 1970 as Katherine Miller Day.

She escorted Gen. Dwight D. Eisenhower to the podium when he was nominated Republican Party candidate; that podium now resides in Hope Farm along with one of the finest collections in the U.S. of early American Empire Furniture, the first pairs of Dorothy Doughty Birds, a six-octave mahogany piano, and many other fine museum pieces.

Mrs. Millers family, who came to Natchez in 1763, descended directly form notables like: Maj. Thomas Grafton, editor of the Natchez Democrat; Thomas Rose, builder of Stanton Hall; Col. Anthony Hutchins, a British officer who settled here in 1772, J.F.H. Claiborne, historian; Benjamine Wade, cotton king; "Light Horse" Harry Lee and Gen. Robert E. Lee.

53. **INGLEWOOD** (1829), this typical southern planter's home was built by Dr. Gustavus Calhoun.

54. **INSTITUTE HALL** (1852-1853), 109 S. Pearl St., was the auditorium for the Natchez Institute, a public school. It has been the Natchez Opera House, the public auditorium, a skating rink, a museum and the site of the first Confederate Pageants.

55. **IVIED OAKS** (1860), 211 S. Rankin. This lovely townhome is a practical part of the history of Natchez as it was built just before the war started.

56. **JEFFERSON MILITARY COLLEGE** (1818), Hwy. 61 N., Washington. Named after President of the United States Thomas Jefferson, who was also President of the American Philosophical Society, this is the oldest Military College in the country, the first educational institution in the territory and the birthplace of Mississippi Statehood.

Complete facilities with nature trail, open to the public from 9:00

AM-5:00 PM daily and from 1:00 PM-5:00 PM on Sunday. 442-2901.

57. **JERSEY SETTLEMENT** (1772), Kingston. In 1772 the Jersey Settlers came to Kingston and in 1773, Anthony Hutchins settled at Coles Creek.

58. **JOHN SMITH HOUSE** (1838), 212 N. Pearl St., Greek Revival Cottage.

59. **JUNKIN HOUSE** S. Rankin. As indicated by its primitive brick structure, this dewlling is very much a part of old Natchez.

60. **KING'S TAVERN** (1789), 613 Jefferson St. Probably the oldest standing house in downtown Natchez and built before 1789, the sturdy stronghold was a stopping place along the Natchez Trace and mail drop for the Trace Post Service. It still has the bullet holes in the door from Indian attack.

61. **LAGONIA** (1787), Hwy 553, Church Hill. This plantation home was built by Marcus Olivares and has stained glass doors. On exhibit are farming tools and a room with original indigo paint, and a restored 1860 roadside doctor's office and barn.

62. **LANSDOWNE** (1853), Pine Ridge Rd., was the first home in Mississippi to have its own gas light plant. The deceptively large and beautiful home sits on the 500 acre wedding present from David Hunt to his daughter Charlotte, who married George Marshall. When Vicksburg fell, a band of Yankee renegades beat and scarred the face of Mrs. Marshall when she tried to stop them from looting the house. Fortunately, a slave had buried most of the valuables under the front portico. The house has always belonged to the same family who built it. It has never been restored, but has remained beautiful through the persistent care of its owners. Bed & Breakfast, 442-3109, 446-9401.

"Lansdowne"

63. **LIBERTY HALL** (1825), 117 S. Pine St. This Federal Style cottage has its front porch wall finished in tongue-and-groove boards which gave it its smooth appearance. This was common for roof-covered portions of the day.

64. **LINDEN** (1818/1849), 1 Linden Place at Melrose Ave., sits on a Spanish land grant to Madame Sara Truly dating 1785, which was bought by Alexander Moore, who's son James built the beginning 4-room, two-story cottage that is the middle part of the house in 1792. Thomas B. Reed, the first U.S. Senator from Mississippi, bought the house in 1818 and called it Reedland. It was Dr. Kerr who changed the name to Linden and sold it to Mrs. Jane Gustine Conner, called the "little war mother" because she sent seven sons and five sons-in-law to war. The Conner family has stayed there for six generations and the home has been filled with priceless family heirlooms over that time. It has a Federal-style doorway that is one of the most famous and photographed of its kind. It was used in the movie "Gone With The Wind".

Open 10:00 AM-3:00 PM daily except during Pilgrimage, Bed & Breakfast year round, AAA 3-Diamond rating, 445-5472.

65. **LONGWOOD** (1861), 140 Lower Woodville Rd. This National Historic

"Linden" Courtesy of Natchez Convention Center.

Landmark has a Byzantine onion-shaped dome on top of a 16-sided copula. The largest octagonal house in the U.S., this fortress of a house, with its 27 inch walls and great dry moat, was never finished by its owner, Dr. Haller Nutt. After spending more than $100,000 and having ordered pieces from all over the world, Dr. Nutt and his family were left to live in the basement rooms, as the northern workers heard the cry of war and fled back to their homes. Later, Dr. Nutt died. His decedents lived in the basement rooms for the next hundred years. The tools and materials and unfulfilled dreams they left are still there.

66. **MAGNOLIA HALL** (1858), 215 S. Pearl. This Greek Revival house

was the last great mansion built in Natchez before the Civil War and has the unique magnolia sculptures throughout. The service wing was hit by a Union cannonball when the Essex fired on the town. It was built by Thomas Henderson, who moved his father's house, called Pleasant Hill, from this site to the corner of Orleans and Pearl. His father was the author of the first book ever printed in the Natchez Territory, was a founding member of the First Presbyterian Church and operated a successful bookstore. Thomas Henderson became a minister of the Presbyterian church down the street and was regarded by several writers as a great and complete Christian.

Now headquarters of the Natchez Garden Club, it is open from 9:00 AM-5:00 PM and has a gift shop and period dress museum. 442-6672. Group meals can be arranged.

67. **McDANIEL TOWNHOUSES** (1852-1853), 600 Jefferson, was built as a rental duplex for Alfred McDaniel.

68. **MELMONT** (1860), 715 N. Rankin St. This pretty home was named using the initials of Mary Elizabeth Lattimore and the hill it rests upon. John Henderson was here when he wrote his appeal for education and religion in the Natchez area.

USS Essex moored at Memphis. Courtesy of United States Army Military History Institute.

69. **MELROSE** (1845), Melrose Ave. John T. McMurran, the law partner of General John Quitman of The Monmouth, built this Palace. McMurran lost his fortune during the Civil War and was killed in a steamboat fire. It was later purchased and revamped in its original style by the Callons of Callon Oil Co. Here in this National Historic Landmark resides the only landscape ever painted by John J. Audubon.

70. **MERCER HOUSE** (1820), 118 S. Wall St. Federal Style town home of William Newton Mercer for his visits to Natchez from Laurel Hill Plantation.

"Melmont" Courtesy of Historic Natchez Foundation.

71. MISTLETOE (1807), Airport Rd. This quiet Mississippi Planters house has belonged to the Bisland and Lambdin families ever since it was built as a wedding present for Peter Bisland and Barbara Foster. It is filled with heirlooms and classic furniture of the period as well as a beautiful garden.

72. MOLASSES FLATS (1805), 200 Main St. Laid down in Flemish bond brickwork, this antique shop now belongs to the author of "Antebellum Natchez", J. Wesley Cooper.

73. MONMOUTH (1818), John A. Quitman Pkwy. This handsome mansion was built by Natchez Postmaster John Hankinson, who died with his wife after helping a yellow fever victim. Later it was the home of Governor, Congressman, and Major General John A. Quitman, who led his troops through Belan Gate, Mexico City and Chapultepec Castle to claim victory in the Mexican-American War where he directed a Natchez boy, Frederick Marcery, to raise Old Glory over the city.

For his services, the U.S. Congress and President James Polk awarded Quitman with a gold sword which is displayed at the house.

Bed & Breakfast is rated by AAA with 4-Diamonds.

74. MONTAIGNE (1855), 200 Liberty Rd. This chalet style home resides among 350 rare varieties of camellias, where William T. Martin once lived with his wife, Margaret Dunlop Conner of Linden. The father of thirteen children, the brave and much loved Mr. Martin served as a Major General in the Confederate Army with J.E.B. Stewart, and held positions as a schoolmaster and district attorney. Union soldiers, freed slaves and scalawags crashed through the lovely home and destroyed much of it priceless interior. One treasure that does remain is the sword presented to General Martin by the Ames Arms Co. of Massachusetts when he reorganized and purchased arms for Adams Troops. The sword was lost in a battle near Shelbyville, TN where he escaped a surrounding force of 1,500 men. The sword was returned ten years later because of the engraving bearing his name.

Though Martin preserved and planted many of the beautiful foliage around the house, it was William Kendall who collected the camellias

"Melrose"

on the grounds. His mother, Mrs. Joseph William Kendall, who started the Coca-Cola Bottling Co. franchise in Natchez in 1906, moved into the house in the 1930's and their descendants remain.

75. **MOUNT LOCUST** (1780), Natchez Trace, was one of the many lodges constructed when the Natchez Trace was in need of rest stops. It remains as a landmark to the time when travel was an unimaginable burden.

76. **MOUNT REPOSE** (1824), Pine Ridge Rd., was built by William Bisland, who's family came here in 1770, and has remained in that family ever since. The Bisland family also built several other houses on the vast estate, including Mount Airwell. Elizabeth Bisland was a talented writer of books and also wrote for the Times Picayune.

77. **MYRTLE BANK** (1836), 408 N. Pearl St. This house was named after the hill that rises above the bluff and serves as the home and Publishing House of Dr. Thomas and Joan Gandy.

78. **MYRTLE TERRACE** (1844-51), 310 N. Pearl St. This was once the home of steamboat Captain Tom Leathers, who raced The Natchez against The Robert E. Lee from New Orleans to St. Louis for $20,000.

In 1870, Captain of the Robert E. Lee took the purse because Leathers stopped in a fog bank for the protection of his passengers.

79. **NATCHEZ UNDER THE HILL** Since this Sub-Colony has contained many different structures from many different periods, we cannot give it a specific date. But La Salle & Tonty were probably the first civilized men to land there. See Chapter 5.

80. **OAKLAND** (Circa 1840), 9 Oak Hurst, was built by Captain Horatio Eustis, who's wife, Catherine Chotard, daughter of the "Yellow Duchess" and Don Estavan Minor, the Englishman who was elected by the Spanish as their Governor of Natchez. This conservative and architecturally unique home of simple beauty, with its 16 1/2 foot ceilings, is sound and timeless. It is now the home of the Lawrence Adams family.

81. **OLD MARYLAND SETTLEMENT** (1770/1810), Hwy 553 off Natchez Trace in Church Hill. This area was initiated by settlers in the early 1770's and Colonel Wood brought many Maryland pioneers to establish a wealthy cotton community. Still remaining are the plantation homes of The Cedars and Lagonia, Christ Church, and the Church Hill Country Store.

82. **THE PARSONAGE** (1852), 305 S. Broadway St. When Peter Little built Rosalie for his wife Eliza, he expected as much of the good life as a palace like Rosalie could offer. However, Eliza came home from school with new religious convictions that brought on the use of the great house as a stop over point for traveling preachers. Peter, not undone by his desire to throw parties, donated the land for the building of The Parsonage around the corner for the comfort and privacy of the moving evangelists.

"Richmond"

84. **PETER CHRIST HOUSE** (1820), 200 S. Pine St. Added to this house in the 1830's was an extra floor. There is a detached kitchen in back of this Federal-style house.

85. **PLEASANT HILL** (1840), 310 S. Pearl St. Moved from the site where Magnolia Hall now stands, this Greek Revival cottage was built for John Henderson, the author of the first book published in the Natchez Territory.

86. **PRESBYTERIAN MANSE** (Between 1824-23), 307 S. Rankin, was built by Benjamine Wade as a symbol of his fortune made in cotton.

87. **PRIEST HOUSE** (Circa 1805), 205 N. Canal St. This house was moved from Market St. and was used as a Law Office.

88. **PROPINQUITY** (1790), off 61 N. In 1797 the first American Political Assembly in the lower Mississippi valley was held here. Later, in 1832, the Benoist family acquired it and it has been in their family ever since. There ancestors planted the many varieties of Camellias and azaleas.

89. **QUEGLES HOUSE** (1830), 508 Washington St., named for the Portuguese family who lived there.

90. **RAVENNA** (Circa 1836), end of S. Union St. Oren and Zuleika Metcalfe, who smuggled food to hungry Confederates during the war, were banished from Natchez until after the war. It was their kinsman, Dr. John Cox, who's life was saved in Timbuktu by the Prince of Jallon. When Dr. Cox stumbled upon the Prince (then a slave) peddling sweet potatoes in Natchez, the Dr. spent the rest of his life trying to free the Prince. His son, William, continued the effort until the Prince was reunited with his whole family in Timbuktu.

"Rosalie"

91. **RICHMOND** (1787/1834/1840), Government Fleet Rd., was built on a 1770 land grant in three connecting stages and in three distinct styles, the first, middle section being more than 200 years old. The house is full of the original furniture and even the piano which accompanied Jenny Lind in her concert in 1851. This intriguing home has been in the Marshall family for 150 years.

92. **RIVERVIEW** (1840/1869). The house was built as a veritable fortress with thirteen inch walls. It and its property have passed through notable owners such as Confederate Lt. George M. Brown, John Steel, acting Governor of the Mississippi Territory, Don Jose Vidal, and Don Estaban Minor.

93. **ROSALIE** (1823), 100 Orleans St. When the Spanish were forced to leave in 1798 and the American flag was raised over Fort Rosalie for the first time, there was only a crude pentagonal stronghold built in a rustic settlement. The American forces soon evacuated the Fort for a newer one near the Spanish border and Fort Rosalie fell into ruin.

A cousin of General George Washington, Col. William Willis, was granted the 22 acre tract of land, which passed to his daughter, Nancy and her husband, Josiah H. McComas. Then it passed to Gamaliel Pease, then to Peter Little. Little owned the first sawmill in the territory and built the present mansion near the ruins of the old Fort Rosalie. The Union soldiers used it as a headquarters. Jefferson Davis visited and General U.S. Grant slept here.

94. **ROUTHLAND** (Circa 1820), 92 Winchester, is filled with priceless furniture and family heirlooms and has belonged to such notables as Job Routh of Dunleith, General and Governor Charles Clark and the

"Stanton Hall"

"Hope Farm" once home to Manuel de Gayoso. Courtesy of Natchez Pilgrimage Tours.

Ratcliffes of Sunnyside Planation to the north. Buried in the cemetery between Dunleith and Routhland is the famous southern writer, Sarah Dorsey. Today the home is still owned by Charles Everette Ratcliffe and his sister, Mrs. Hector H. Howard, who's home is Rip Rap.

95. **ST. MARY'S CATHEDRAL** (1842), 107 S. Union St., is obviously one of the most glorious places of worship in the south. It took many years to complete and was consecrated in 1886 only after it was free from debt.

96. **SAMUEL COCKERELL HOUSE** (1830/1850), 207 S. Union St. As evidenced by the name on the gate, this was the house of silversmith, Samuel Cockerell.

97. **SARAGOSSA** (before 1825), Sara Gosa Rd., Just off the "Old Flatboatman's Road" which extended the Natchez Trace to New Orleans, this West Indian influenced plantation house. It was owned by many notables such as Colonel Winthrop Sargent, the first Territorial Governor Of Mississippi, Jonathan Thompson, Dr. Stephen Duncan of Auburn, William St. John Elliot, the builder of D'Evereux, the distinguished Austin Williams and his wife, formerly Caroline Matelda Routh, and Walton Pembroke Smith. For the most part it has stayed in the Smith family since.

Walton Pembroke Smith was the son of William Haslet Smith and his wife, Mary Bell Madison, the daughter of Francis and Susan Bell Madison, the only niece of President James Madison. Walton Smith was "a man of brilliant education and striking personality" who used the compass that remains at Saragossa to survey his properties and started to survey the Mississippi River and its tributaries. His son, Austin Williams Smith, completed that survey and published the pamphlet proposing the famous Tennessee-Tombigby canal.

89. **SCOTT HOME** (1796), N. Pearl St. This home was also given by Andrew Marschalk to his youngest daughter, Jane Elenore.

99. **SELMA** (1815), off 61 N. This was an Indigo plantation built by the first Native-borne Governor of Mississippi, Gerard Brandon.

100. **SHIELDS TOWNHOUSE** (Circa 1860), 701 N. Union, was the last surviving townhouse built before the Civil War.

101. **SPRINGFIELD** (1790), Built before 1790 and the marriage place of Andrew Jackson to Rachel Donelson Robards, it was the home of Magistrate Thomas Green.

102. **STANTON HALL** (1857), 401 High St. From his immense fortune made as a cotton broker, Frederick Stanton had the best of everything sent from overseas to decorate this National Historic Landmark he called "Belfast" for his Irish beginnings. It cost the overblown price of nearly $90,000. It has 16 1/2 foot ceilings, silver door hardware, bronze, gas-burning chandeliers (Gasoliers) especially made for the rooms they occupy, the finest white Carrara marble mantels, specially made French mirrors, a oversized antique Aubusson carpet, Old Paris china, and an elliptical stairway with mahogany rail. At the turn of the century the house was used as the Stanton College for Young Ladies. The Pilgrimage Garden Club now uses it for their headquarters.

103. **TEXADA** (Circa 1790), 222 S. Wall St. At one time or another it has

The Steamboat "Natchez" unloading supplies to the Confederate Army at Baton Rouge. Courtesy United States Army Military History Institute.

been used as many things: Tavern, dance academy, tailor shop, elephant exhibition, Antique Shop, and even the house of Don Manuel de Texada.

104. THE TOWERS (Circa 1820), 801 Myrtle This Italian design home stands compelling and unique in its magnolia-framed setting. The front of it was constructed around 1859.

105. TILLMAN HOUSE (1837), 506 High St. This house was remodeled in 1860. It has been a girls school and Dr. Dub's dental office.

106. TRINITY EPISCOPAL CHURCH (1822), 305 S. Commerce, was Federal style and had Greek Revival front added in 1838, and the Romanesque parish house was added in 1886 by famous architect Frederick C. Withers.

107. TWIN OAKS (1812), 71 Homochitto St. Built by Jonathan Thompson, for territorial sheriff and cotton planter, Lewis Evans, this lovely house has been occupied by the Connellys who later did great works in the Catholic clergy. Pierce Connelly was at one time the rector of Trinity Episcopal Church before he took his wife to Rome to join the order. His wife, Mother Cornelia Connelly, who called the home White Cottage, was sent to England by the Pope and founded the Society Of Holy Child Jesus for the teaching of young girls. There is a chapel in one of the gardens with Tiffany stained glass windows to honor her.

The house has been owned by three doctors and Charles L. Dubuisson,

"The Towers" Courtesy of Bill Stewart.

who was president of Jefferson College, a judge and a representative of the Natchez area in the State Legislature. Then the Whittingtons changed the name because it was repainted a different color after a fire. The Whittington family has served this country's military since the Revolutionary War.

In the house are found many fine porcelains, furniture, and paintings, including a double portrait by M. Petit, a court painter of Marie Antoinette, a mountain landscape by Wesler, and a painting of the Natchez Trace by Sharpe. During refurbishment by the Whittingtons, it was discovered that the door hardware is made of fine Sheffield Silver.

108. VAN COURT TOWNHOUSE (1835), 510 Washington St. This tall, elegant house displays the combined features of Greek and Federal periods.

109. WALSH HOUSE (1839), 216 S. Union St. As originally built, like a condominium, this town house shares a wall with its neighbor.

110. WALWORTH LAW OFFICE (1840), 116 S. Wall St. The owner of The Burn, John P. Walworth, used this building as his law office.

111. WEYMOUTH HALL (1855), 1 Old Cemetery Rd., is situated on top of the bluff with a magnificent, panoramic view of the Mississippi River. Winner of the Mississippi Historical Society's 1985 Preservation Award, this mansion was occupied by Union soldiers during the war. It has an extensive collection of period antiques and is open for tour and Bed & Breakfast. 445-2304.

112. WHITE PILLARS (1859), 206 S. Pine St. This majestic town home was built by Mrs. Missouri Lawrence and is now owned by Lucius Butts.

113. WHITE TURPIN HOUSE (1818), 608 Jefferson St., was built by the Sheriff of Adams County, White Turpin.

114. WHITE WINGS (1833/1854/1892) 311 N. Wall St. Added to during several periods, this unique home speaks of those casual southern summers.

115. WIGWAM (1836), 307 Oak St. The original section was built before 1790 and renovated in 1839. It is believed that it acquired its name because it was built upon an Indian Burial Ground. Artifacts are very often found around it.

116. WILKINS HOUSE (1830), 300 N. Commerce St. This cottage sports brick-linked chimneys.

117. WILLIAM AILES HOUSE (1853), this fine home overlooking the Mississippi River was part of Belleview Plantation. The old home called Belleview no longer exists.

118. WILLIAM HARRIS HOUSE (1835), 311 Jefferson St. This double house, because it was originally constructed of brick (the 1st floor) and wood (2nd flood), was built by the father of Confederate General Nathaniel Harris.

119. WILLIAMSON HOUSE (Circa 1850), 208 S. Pine. This house was built of brick and New Orleans style iron grille work.

120. ZION CHAPEL AFRICAN METHODIST EPISCOPAL CHURCH (1858), 228 Pine St., was originally built as the Second Presbyterian Church. Minister of the African Methodist Episcopal Church, Hiram R. Revels, was the first black to ever serve in either house of congress.

II

MODERN NATCHEZ

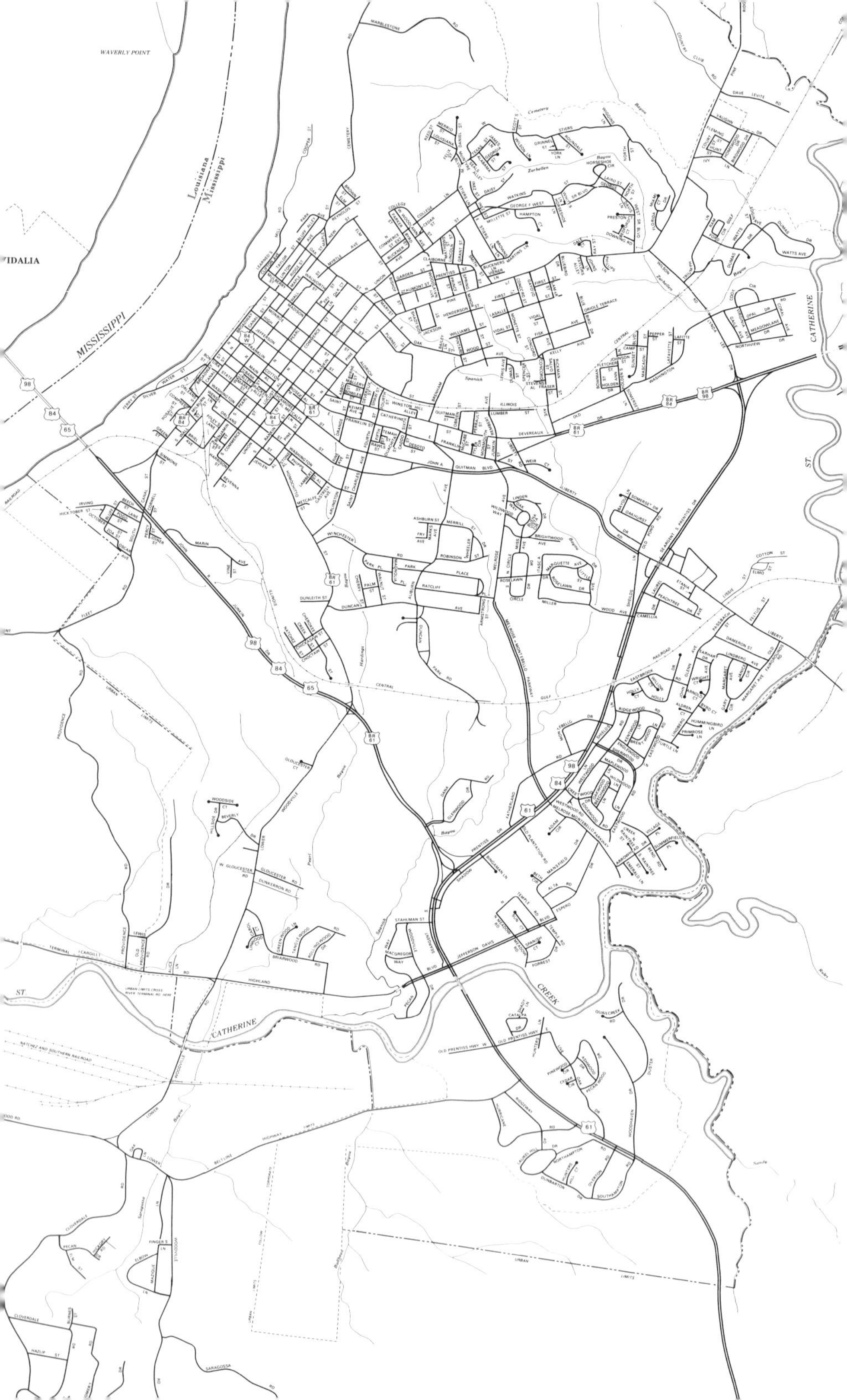

WAVERLY POINT
VIDALIA
Louisiana
Mississippi
MISSISSIPPI
CATHERINE
ST.
CREEK
CATHERINE
ST.
98
84
65
BR 61
BR 84
BR 98
61
MARBLESTONE RD
Cemetery
Bayou
Zurhellen
Spanish
Hardings
Pearl
CENTRAL
GULF
RAILROAD
ILLINOIS
URBAN LIMITS
JOHN A QUITMAN BLVD
MELROSE - MONTEBELLO PARKWAY
SERGEANT S. PRENTISS DR
LIBERTY
JEFFERSON DAVIS BLVD
HIGHLAND
TERMINAL (CARGILL)
NATCHEZ AND SOUTHERN RAILROAD
HIGHWAY
BELT LINE
OLD PRENTISS HWY
SARAGOSSA
CLOVERDALE
HAZLIP ST
LOWER WOODVILLE
DUNCAN AVE
RATCLIFF
PLACE
ROSELAWN
MILLER
WOOD AVE
WINCHESTER
DUNLEITH ST
KELLY AVE
WASHINGTON
DEVEREAUX
LUMBER
COUNTRY CLUB
DAVE LEVITE RD
WATTS AVE
DUMAS

3

THE COMMUNITY

Natchez is a beautiful city with a vibrant economy. Relying mostly on the tourist trade, the people of Natchez do everything possible to make the community attractive and receptive to its visitors. It is kept clean and groomed and the streets and highways are developed to their fullest efficiency. This makes the visit to Natchez pleasant and relaxing and without ordeal. It follows that our tourists almost always want to come back.

As for the newcomer, Natchez is accommodating and helpful to anyone who transfers or relocates within her benign influence. It is easy to find the necessary places of business to get the settler started, and there is a Welcome Center at 370 Seargent S. Prentiss Dr., (601) 442-5849, which gives free welcome packages and helpful information. There is also the Natchez-Adams County Chamber Of Commerce, 205 N. Canal St., (601) 445-4611; they are always willing to help the new citizen to set up and get started. Natchez wants her people to prosper.

STATISTICS

At an Altitude of 218 feet above sea level, the 448 square mile area that makes up Adams County and the Natchez area has mild winters, warm summers at an average low of 55° and an average high of 75°. Average annual rainfall is 55 inches, with a growing season of 250 days between March 10th and November 15th. The city has an estimated population of 22,500 and the county, 39,000 with a trade area of 100,000 people.

Compared to other cities the same general size as Natchez, her central business district shows a much stronger stature. The Chamber of Commerce has nearly 800 member businesses. The county also works hard to make its trade successful; where in 1984, with a $238.4 million retail sales volume, her population is projected to increase by 1995 to 40,200 with a sales volume of $336 million, and personal income will be at $611 million. To bring the per capita income to $15,200.

RESOURCES

AGRICULTURE: Cotton, Soybeans, Corn, Rice, Peanuts, Wheat, Watermelon, Catfish, Timber, Gravel, Stone, Etc.

OIL: Around 50 Oil Producing Companies operate in this area and dozens more supporting companies.

MANUFACTURING: At least 18 manufacturers from paper to tires, petroleum products to frozen foods, lumber to soft drinks, sand, gravel and concrete to meats and ice.

TOURISM: Well over 150,000 tourists yearly, and it's increasing.

COMMUNITY SERVICES, INSTITUTIONS & FACILITIES

Below, in alphabetical order, is a listing of services and facilities available. Further listings of the most needed places of business are presented in other chapters (quick reference in the index). Even though a complete list would be impossible in a guide this size, there is enough information for the average visitor to get around without much more effort than reading it here. Any other information can be obtained from the Natchez Pilgrimage Tour Headquarters, the Chamber of Commerce or the Welcome Center--all listed below.

ADAMS COUNTY FARM BUREAU: Liberty Rd. 445-4187.

ADAMS COUNTY FORESTRY DEPT.: Liberty Rd. 442-0472.

ADAMS COUNTY SOIL CONSERVATION: 329 Liberty Rd. 442-5351.

AGRICULTURAL COUNTY AGENT: Liberty Rd. 445-8201.

ARCHIVES & HISTORY DEPT.: Contact Grand Village of the Natchez Indians or Jefferson College listed in this chapter.

BABY-SITTING SERVICES: Listed in the yellow pages of the phone book under Child Care Centers is a long list of day-care services available for about $6.00 a day.

CAMERA SHOPS: For One Hour film processing see Photography below.

CHAMBER OF COMMERCE: 205 N. Canal St. 445-4611. The Chamber offers publications on the Natchez economy, maps, mailing lists of the member businesses, directories and videos.

CHURCHES: Natchez has some 92 churches under 25 denominations. (See Yellow Pages of the telephone directory).

CIVIC ORGANIZATIONS: American Legion, Colonial Dames, Daughters Of The American Revolution, Elks, Jaycees, Junior Auxiliary, Kiwanis, Lions, Moose, Rotary, Shriners, United Daughters Of the Confederacy. Contact the Chamber of Commerce.

COMMUNICATIONS:

NEWSPAPER

The Natchez Democrat, mornings except Monday, 503 N. Canal St. 442-9101.

RADIO

K-10-AM (1040) Adult rock and weather, Beltline Rd. 446-8803

WMIS-AM-NBC (1240) The soul of Natchez, 20 E. Franklin 442-2522.

WNAT-AM-ABC (1450) Miss-Lou's Award winning station, 2 O'Ferrall 442-4895.

WQNZ-FM-ABC (95.1) 100,000 watts stereo country, 2 O'Ferrall 442-4895.

WTYJ-FM (98) Rock, 20 E. Franklin St. 442-2522.
KFNV-FM (107.1) Rock, Vidalia Hwy., Ferriday 757-4200.
KFNV-AM (1600) Vidalia Hwy., Ferriday 757-4200
KQID-FM (93) 1115 Texas Ave., Alexandria, LA 336-7625.
KNOE-FM-ABC (102) P.O. Box 4067, Monroe, LA 71211. Request 387-8102.
KNOE-AM-ABC (540) same as above. Offices 388-8888.

"St. Mary's Cathedral" Courtesy Mississippi Dept. of Archives & History.

TELEVISION

For cable hook-up and pay T.V. in Natchez call Sammons-Communications, 190 Hwy. 61 S. 442-5418, 24 Hour Repair 442-0368.

Channel	Call Sign		
2	WBRZ	(BatonRouge)	ABC
3	WLBT	(Jackson)	NBC
4	WTBS	(Atlanta)	TBS
5	KALB	(Alexandria)	NBC
6	WLBT	(Jackson)	NBC
7	WMAU	(Bude)	PBS
8	KNOE	(Monroe)	CBS
9	CNN	(Atlanta)	CNN
10	ESPN	(Sports)	ESPN
11	WAPT	(Jackson)	ABC
12	WJTV	(Jackson)	CBS
13	CBN		CBN

CONVENTION CENTER: Conveniently located at Liberty Park only a mile and a half from the Mississippi River, the 18,000 square foot Natchez Convention Center provides complete modern facilities to seat up to 2,000 people auditorium style, 1,300 banquet style, and can hold up to 86

8'x10' booths. There is a large truck entrance, an available office, a fully equipped kitchen, several smaller meeting rooms. The friendly people at the Center can customize your meeting needs and can arrange tours, receptions and parties at one of the magnificent antebellum homes. They offer complete information on special and annual events. Here also is the Film Commission which scouts locations for the movie industry. Located off Liberty road behind the ball park. 1 (800) 647-6724, 446-6345.

COURTHOUSE: County Judge, 442-6416.

EVENTS: Besides Christmas in Natchez and the Spring and Fall Pilgrimages, there are dozens of other things happening in Natchez throughout the year, like the Balloon race, Antique shows, Sporting events and Christmas In Natchez. See Events Calendar in Chapter 8.

EDUCATIONAL FACILITIES:

Natchez-Adams County Public Schools, 442-0212.
Adams County Christian School System, 442-1425.
Cathedral School, 442-1988.
First United Pentecostal Church, 442-1411.
Holy Family, 442-3947.
Maranatha Baptist School, 442-7385.
Natchez Christian Academy, 442-1411.
Pleasant Acre Day School, 442-2264.
Trinity Day School, 442-4756.
Alcorn State University Of Nursing, 442-3901.
Copiah-Lincoln Junior College Natchez, 442-9111.
Natchez College, 445-9702.

FARMER'S HOME ADMINISTRATION: If you are buying a home for the first time or need help with property financing, you must contact them at 329 Liberty Rd. 442-5351.

FILM COMMISSION: Since Natchez has been a prime spot for filming movies, the Convention Center has a Film Commission to help those in the industry scout locations. 1 (800) 647-6724, 446-6345.

FINANCIAL INSTITUTIONS:

Banks

Britton & Koontz First National Bank, 445-5576
Deposit Guaranty National Bank, 442-6471.
Magnolia Federal Bank, 442-2772.
United Mississippi Bank, 446-7761.

Savings & Loans

Magnolia Federal, 442-2772.
Natchez First Federal Savings & Loan, 442-2733.

FISH & GAME WARDEN: Mississippi Dept. of Wildlife Conservation, P.O. Box 451, Jackson, MS 39205, 961-5300. For hunting and fishing seasons see Hunting or Fishing in Recreation Chapter 6.

GOVERNMENT:
Mayor-Board Of Aldermen, City Hall, 446-6641.
County Board Of Supervisors, Court House, 442-2431.

HEALTH COUNSELLING & FAMILY PLANNING: 408 Hwy. 61 N. 446-7940.

Firefighting in Old Natchez. Courtesy of Bill Stewart.

HISTORIC NATCHEZ FOUNDATION: 109 N. Commerce St. 442-2500.

HOSPITALS:
Jefferson Davis Memorial Hospital, Sgt. S. Prentiss Dr., (206 beds), 442-2871.
Humana Hospital Natchez, Jefferson Davis Blvd., (101 beds), 446-7711.

LAUNDRIES & LAUNDROMATS:
Allen's Trace Town Cleaners, Trace Town Shopping Ctr. 445-8999.
Drive In Cleaners, 193 Homochitto St. 445-8294.
Grace Cleaners, 30 Sgt. S. Prentiss Dr. 442-8420.
Natchez Steam Laundry, 123 St. Catherine St. 445-5668.

LAW ENFORCEMENT:
Police Department, 445-5565.
Sheriff, 442-2752.
Fire Department, 442-7373.

LAWS & LICENSES: New Residents have 30 days to purchase license plates from County Tax Collector, 442-8601; 60 days for 15 year-old's or over to get drivers license at Highway Patrol Office, 442-4897. –Motor Vehicle Inspection is required. –Homestead Exemption must be made by April 1st at County Court House. –State Income Tax is due April 15th to state Tax Commission. Voters must be 18 or over, a U.S. citizen and must be registered in this district 30 days (Court House) prior to election. –Marriage Applications ($20) are made at County Circuit Clerk's office in the Court House; issued three days after application and a blood test is required.

LIBRARY: Judge George W. Armstrong Library, Containing over 70,000 books, the free facility is Open Monday - Thursday, 9:00 A.M.- 6:00 P.M., Friday, 9:00 A.M.- 5:00 P.M., Saturday, 9:00 A.M.-1:00 P.M. Closed during Christmas. Commerce at Washington St. 445-8862.

LOCKSMITHS:

Power Safe & Lock Service, 101 Carter St., Vidalia 336-5288.
Wayne Lock & Key Service, 240 D'Evereux Dr. 442-1063, 442-4433.

MAPS: Can be obtained from the Chamber Of Commerce.

Cannon Practice on the Bluff. Courtesy of Bill Stewart.

MOTELS: 12, containing 1,000 rooms and 23 Bed & Breakfast houses with over 70 rooms. See Lodging in Chapter 4.

PARKS: See Camping in Recreation Chapter 6. Natchez National Historic Park will be located in downtown Natchez, stretching along the river bluff. It is to be the terminating point of the Natchez Trace Parkway.
Natchez State Park, at Stanton, 442-2658.
Natchez Trace Parkway, off Hwy. 61 N. 446-4211.

POSTAL SERVICES:

Federal Express Box, Brown Building, 521 Main St. at the end of the hall.
Natchez Post Office, 214 N. Canal St. 442-4361.
MS Mail Savings Inc., 118 Jefferson Davis Blvd. 442-0042.
Sibley Post Office, Hwy. 61 S. 442-6006.
Tracetown Post Office, 4 Tracetown Shopping Center, 442-5518.
Washington Post Office, Hwy. 61 N. 442-5671.
United Parcel Service, Highland Blvd. 1 (800) 228-8088.

RECREATIONAL FACILITIES: Natchez Recreation Department, 442-7616.

SECRETARIAL SERVICES: A-1 Reporting & Secretarial Service, 446-6179.

TOUR INFORMATION CENTER: Natchez Pilgrimage Tours, Canal, 446-6631.

TRANSPORTATION: See Chapter 4. Air traffic is handled by the Natchez-Adams County Airport with general facilities and 6,500 feet of lighted, hard surface runway. –Railroad shipping is handled by the Western Railroad Association. –Bus Service is provided by Trailways and Greyhound. –Highways are 61, 84, and 98. City streets are 100% paved.

The Old Natchez Hotel. Courtesy Mississippi Dept. of Archives & History.

UTILITIES:

Electricity

Mississippi Power & Light, 247 D'Evereux Dr. 442-7474.
Southwest Mississippi Electric Power Ass., Lorman, MS 442-2493.

Gas

Mid-Louisiana Gas Company, Hwy 61 S. 442-6491.
Mississippi Valley Gas Company, 54 E. Franklin St. 442-4831.

Telephone

South Central Bell, D'Evereux Dr. 442-9011. Area Code for all of Mississippi: 601.

Water

Natchez Water Works, 307 Market St. 445-5521.

VETERINARIANS:

Bluff City Veterinary Service, 422 Hwy. 61 N. 446-5899.
Borum's Veterinary Clinic, 145 E. Franklin St. 442-5522.
Dunkerron Animal Hospital, 612 Highland Park Blvd. 442-7002.
Galbreth, Dr. Scott, 508 Concord St. 445-9924.
Natchez Veterinary Clinic, 404 Liberty Rd. 445-5271.

WELCOME CENTER: 370 Seargent S. Prentiss Dr. 442-5849

WESTERN UNION: To pick up or send money or messages–the bus depot, 103 Lower Woodville Rd. at John R. Junkin Dr. 445-5291. Mailgram, Telegram, or Cablegram--1 (800) 325-6000.

Antebellum Town House "Stanton Hall" Photo by Mark Coffee

4

SEEING NATCHEZ

Whether you drive, walk, cycle, tour or carriage through Natchez, you must take a long, close look at its architecture and its people. Both are decedents of a different sort of culture, compiled from different nations of the world. All this blends together to form an unusual society very much set apart from the rest of the world. You will find a little something for everyone whether it be architecture, sports, art, theater or nature. Stop and study closely everything that interests you, and you might see into that past which brought us this lovely bit of living history.

Thanks to Spain's Manuel de Gayoso, the streets of downtown Natchez are comfortably organized in square blocks, so finding your way around this part of town will be easy. However, as the streets spread out into the countryside, reckoning becomes more and more complicated. Because of the hills that make up the loess bluff, snaking paths were easier to follow and roads more often found their beds laid on the same paths, which led to the plantations, the Natchez Trace,

Filming "Beulah Land" at Grace Methodist Church on Pearl St. © 1980 Columbia Pictures Television, A division of Columbia Pictures Industries, Inc. Courtesy of Columbia Pictures. All Rights Reserved

and the outer settlements. But don't let the complexity of country travel shy you away from one of the most scenic settings in existence. For your convenience, there is a walking guide map showing most of the points of interest in downtown Natchez at the beginning of the next chapter.

Many reference points serve the visitor in finding their way around. The Mississippi River bridges, just south if the downtown area, connecting Vidalia, Louisiana and Natchez are part of Highways 98, 84, and 65. This same four-lane highway, John R. Junkin Dr. or the "By-Pass", leads around the outskirts to exit Natchez to the south and northeast via Seargent S. Prentiss Dr. To depart town is to take one of these three exits. The Mississippi river bridges lead to anywhere in Louisiana except Baton Rouge, New Orleans, St. Francisville and anything else east of the Mississippi. These towns are accessed by U.S. Highways 61 and 65, which lead south past the Jefferson Davis Memorial Hospital. Connecting all the highways to the north and east 61, 84, 98, and the Natchez Trace Parkway are two main roads: Seargent S. Prentiss Dr., which connects north to south and north to west; and D'evereux Dr., which connects north to downtown. To leave downtown headed north or east, take Franklin St. at the river and head east, away from the river. Franklin will veer left onto D'evereux Dr. just past the Texaco at the Shell station on Melrose and connect with The Natchez Trace, and highways 61 north, and 84 & 98 east. Pay close attention to the one-way streets, as some of them change in mid-stream.

During their stay, people have gotten around the Natchez area in many ways: from Ballooning to Horseback, from flatboat and paddle wheelers to airboats and jet skis. Though some of these modes are impractical, some of the elegant old transports are still used. Natchez still has horse-drawn carriages as listed below. And, though rarely spotted, ballooning has been seen here since they were contrived. Paddle wheelers like the "Mississippi Queen" and the "Delta Queen" still visit Natchez once a week.

Practically, though, Natchez is a tourist town. And there is a private transit system that offers excellent tours of the homes, the downtown area and the

Streetcars pulled by mule through downtown Natchez. Courtesy Prince & Natchez Photo Lab.

countryside. Tour busses stop at most of the hotels daily. You can arrange reservations or set up a custom tour. Whatever your preferred mode of transportation, you will probably find it listed below.

TRANSPORTATION

AIR TRAVEL:

Airport (Operations) 442-3142, (Manager) 442-5171. Daily flights to Jackson and New Orleans.
Alpine Flying Service Jonesville, La 339-8722.
Flight Line Inc. 1 (800) 223-1891.
Silver Mount Inc. 603 Franklin St. 445-2150, 1 (800) 423-0684.
South Central Group Inc. Canal Street Depot 442-1603.
Travel Agencies also provide Airline Tickets

AUTOMOBILE RENTING:

Adams County Car Rental & Taxi Cab Co. Airport Rd. 445-9831, 446-5454.
Add-A-Car Rental 1402 Carter St. Vidalia 336-5380.
Buick Dealer Renting and Leasing Williams Motor Co. Hwy 84 Ferriday, LA (318) 757-9211.
Community Ford Vidalia Hwy Ferriday, LA 442-7251.
Dossett Auto World Hwy. 61 N. 442-4301, 446-5800.
Ford Dealer Leasing and Renting Trace City Ford Hwy 61 N. 446-5880.
Natchez Rent A Car Inc. Seargent S. Prentiss Dr. 442-0764.
Ugly Duckling Rent-A-Car 903 Carter St. Vidalia, LA 336-7144.

BICYCLES:

Natchez Bicycling Center 343 Main St. 446-7794.
Coast To Coast 196 Seargent S. Prentiss Dr. 442-4352.
Western Auto Associate Store 601 Main St. 445-4186.

BOAT RENTING and LEASING:

Lonnie's Canoe Rental 605 Hwy 49 N. Seminary 722-4301.
Tate S J Boat Camp Lake Concordia Ferriday, LA 757-9939.

BUS LINES:

Bus Station 202 La. Hwy Ferriday, LA 757-8196.
Dixieland Tours 446-6311.
Continental Trailways and Ramsey 445-8225.
Greyhound Trailways Bus Lines 103 Lower Woodville Rd. 445-5291.
Hotard Coaches 442-0876.
Coaches Inc. 1 (800) 233-1219, 1 (800) 553-4895.
Natchez Transit System 800 Washington St. 442-9909.

CARRIAGE RIDES:

Three Carriage companies charge about $5 per person and are available at

the Canal St. Depot and can pick-up at most downtown addresses. 446-6631.
Magnolia Carriage Co. 446-6631.
Natchez Carriage Co. 446-6631.
Southern Carriage Co. 442-2151, 1 (800) 647-6742.

TOUR GROUPS:

Natchez Pilgrimage Tours Canal St. Depot 446-6631. See Tours Below.

RIVER TRAVEL

American Cruise Lines for the Savannah, 1 (800) 243-6755.
Delta Queen Steamboat Company 1 (800) 543-7637.
Mississippi Queen stops at Silver Street Wednesday and Delta Queen occasionally

TRAVEL AGENCIES:

Natchez Travel Service (American Express Representative) 118 N. Pearl St. 442-6001.
Creative Travel Canal St. Depot 442-3762, 1 (800) 824-0355.
United Mississippi Bank (Travelers' Checks Issued) 446-7761.

WALKING:

Natchez Walking Guide Book Available through The Historic Natchez Foundation.

"The Burn" Courtesy of Natchez Convention Center.

TOURS

Since the 1930's Natchez has been a world-wide interest as tourists have come to see the old mansions from the bygone days of colonization and the civil war. Today the homes live as a legend to those interested in architecture and history; they bathe in splendor for those interested in beauty; and they testify to the culture, excellence, and agelessness of the spirit.

Visit these palatial monuments, surrounded in tranquil gardens and filled into every corner with history and beauty. Lovely southern bells in hoop skirts guide you through and tell you of this history so enticing to the imagination.

Natchez Pilgrimage Tours offers the 1989 tours below. The same tours in other years will be off a day or two from this publishing, but free brochures are available from the tour companies. Spring Pilgrimage is from March 11 - April 9, 1989, March 10 - April 8, 1990, and March 9 - April 7, 1991. The Fall Pilgrimage dates are: October 7 - 21, 1989, October 6 - 20, 1990. There are morning and afternoon tours which cover most of the homes open to the public. Larger Group Reservations and rates, Private Guides, Motor Coach transportation and Half-price tickets for children (ages 6-17) are available. Tickets can only be obtained from Natchez Pilgrimage Tours; no tickets are sold at individual houses. Visa, Mater Card and American Express accepted. Below is a listing of the tours held which are numbered to coincide with the descriptions in Chapter 2 so descriptions of the homes can be found. Also listed are prices and times. For information, reservations and tickets contact:

Natchez Pilgrimage Tours

(Open Daily 8:30 A.M. - 5:30 P.M.)
P.O. Box 347 * Canal at State Street
Natchez, MS 39121
(601) 446-6631 * 1-800-647-6742

"The Parsonage" long ago. Courtesy of Bill Stewart.

TOURS:

Morning Tour 9:30 A.M.-1:00 P.M.
Afternoon Tour 2:00 P.M.-5:30 P.M.

PRICES:

3 House Tour $12 per person
4 House Tour $14 per person
5 House Tour $16 per person

TOUR TIMES

Orange - Morning Tour 9:30 A.M. - 1:00 P.M.

March 11, 14 ,17, 20, 23, 26, 29 April 1, 4, 7
7 The Banker's House, 107 S. Canal
107 Twin Oaks, 71 Homochitto St.
65 Longwood, Lower Woodville Rd.
49 Hawthorne, Lower Woodville Rd.
33 Elms Court, John R. Junkin Dr.

Green - Afternoon Tour 2:00 P.M. - 5:30 P.M.

March 11, 14, 17, 20, 23, 26, 29, April 1, 4, 7
73 Monmouth, John A. Quitman Pkwy.
74 Montaigne, 200 Liberty Rd.
80 Oakland, Oakhurst off Old Pond Rd.
26 D'Evereux, D'Evereux Dr.
62 Lansdowne, Pine Ridge Rd.

Red Tour - Morning Tour 9:30 A.M. - 1:00 P.M.

March 12, 15, 18, 21, 24, 27, 30, April 2, 5, 8
22 Connelly's Tavern, N. Canal at Jefferson
17 Cherokee, High at Wall St.
32 The Elms, S. Pine at Washington
64 Linden, Melrose Ave.
69 Melrose, Melrose Ave.

Yellow - Afternoon Tour 2:00 P.M. - 5:30 P.M.

March 12, 15, 18, 21, 24, 27, 30, April 2, 5, 8
46 Green Leaves, S. Rankin at Washington St.
29 Dunleith, 84 Homochitto St.
31 Elgin, Elgin R. off Hwy. 61 S.
36 Fair Oaks, Hwy. 61 S.
102 Stanton Hall, High at Pearl St.

Blue- Morning Tour 9:30 A.M. - 1:00 P.M.

March 13, 16, 19, 22, 25, 28, 31 April 3, 6, 9
15 The Burn, 712 N. Union St.
30 Edgewood, Airport Rd.
71 Mistletoe, Airport Rd.
76 Mount Repose, Pine Ridge Rd.
66 Magnolia Hall, Pearl St. at Washington

Pink - Afternoon Tour 2:00 P.M. - 5:30 P.M.

March 13, 16, 19, 22, 25, 28, 31 April 3, 6, 9
93 Rosalie, S. Broadway at Canal St.
82 The Parsonage, 205 S. Broadway St.
91 Richmond, Gvt. Fleet Rd.
52 Hope Farm, Homochitto at Duncan Ave.
94 Routhland, 92 Winchester Rd.

LODGING

"Weymouth Hall" Photo by Andy Fisher.

Bed & Breakfast

To stay in one of these historic rooms is to take a relaxing trip back in time. You can feel the way those elegant ladies and gentlemen of the Old South lived, the simplicity and spirituality that graced those bygone days of leisurely enjoyment. A unique and compelling experience, the Bed & Breakfast can add a nostalgic moment and a fond memory to your life.

Be a part of those charming and noble plantation dwellers in one of over 70 antebellum rooms Natchez offers. Some of the great houses have 13 rooms to chose from, each with some alluring fascination that is unique to Natchez and tempting to the imagination.

Some of the amenities offered are listed below, but further information on the history of the homes can be found in Chapter 2. While some of the homes are only open for tour during the Pilgrimage, most are open for Bed & Breakfast all year round. As with nearly all Bed & Breakfasts, reservations are needed, at least during the Pilgrimage season. The younger crowd and those not interested in the Pilgrimage might prefer to stay in summer or winter. At any rate, if you are it town, you might be able to find accommodations by calling some of them.

Bed & Breakfast always comes with a complimentary breakfast, housekeeping and tour of the home. While there are sometimes restrictions on age or pets or property access, it should be understood that the host's private quarters are restricted. The tour and stay is enough to get a feeling of how the southerners of the past lived. It is practically overwhelming.

Smoking is rarely allowed because of its effect on the antique furniture. Some offer party and reception facilities and others offer some sort of recreation, like fishing or evening cocktails. No matter what each home has to offer in the way of private luxuries, the Natchez Bed & Breakfast is a wonderful experience and well worth living.

Natchez Pilgrimage Tours handles Bed & Breakfast reservations and information for most of the homes. Reservations during Pilgrimage should be made 2 to 3 months in advance to get desired placement and a deposit is necessary. However, it is still possible to walk in. For a stay of one night, payment is made in advance. For more than one night, deposits are required. Prices range from $50-$150. Ask For Bed & Breakfast Department, 1-800-647-6742 or (601) 446-6631.

AUBURN (1812), 400 Duncan Ave. 442-5981. This National Historic Landmark is now part of a graceful park, complete with pool, tennis courts and golf course all right out the front door. Open all year from 9:00-5:00 except Christmas, Auburn has 4 rooms for 2 with Southern breakfast complimentary. Each has a private bath. They prefer smoking done outdoors but offer a free tour and handicap ramp. There are no restrictions on age but reservations with a deposit is required.

AUNT CLARA'S COTTAGE 712 N. Union, part of The Burn 442-1344.

THE BURN (1835), 712 N. Union St. 442-1344. Open all year round from 9:00-5:00 except Christmas, The Burn has 10 rooms for 2. A Plantation Breakfast of hot grits, country ham, eggs, fresh biscuits, sausage, toast, fruit and piping hot coffee come with the $60-$125 price as well as a complimentary drink on arrival. The rooms have private bath and T.V. Smoking is allowed. Access to the whole garden property and swimming pool are provided. There are no age restrictions. Dinner parties, picnics are available and catering for groups costs $26.50 or $32.00 per person. Visa/MasterCard accepted.

D'EVEREUX (1840), D'Evereux Dr. 446-8169. Open all year, this magnificent home has 1 room for 2 at $100. Smoking is preferred outside and a free tour is included along with breakfast, private bath and phone. Children 12 or over are welcome, but no pets. Access to the expanse of beautiful property is also included and reservations are needed.

DUNLEITH (1856), 84 Homochitto St. 446-8500, 446-6094. Open for tour all year from 9:00-5:00 and 12:30-5:00 on Sundays except holidays. This National Historic Landmark has 11 rooms for 2 or 3 people ranging from $85-$130. Free tour, private bath and 40 acres of manicured gardens and grounds accompany this magnificent manor home. Also included are private phones, bath, T.V., and a porch that wraps around the house. Smoking is allowed. Reservations are needed. Visa/MasterCard/American Express/Discover Card are accepted, and banquets, receptions, parties and catering are available.

ELGIN (1800), Hwy. 61 S. in Second Creek Neighborhood 446-6100. 3 rooms for 2 are open during Spring Pilgrimage in a lovely dependency on beautifully flowered grounds. With access to this old plantation home comes a full Plantation Breakfast, private bath and T.V. Smoking is allowed. Visa/MasterCard.

GUESTHOUSE IN NATCHEZ (1840) 201 N. Pearl St. 442-1054. Open 24 hours all year round, this downtown extension of the Eola Hotel has 18 rooms for 2 starting at $75. Handicap ramp and bathroom rails, private

"Elgin" Plantation Bed & Breakfast. Courtesy the Calhoun Family.

phone, bath, and T.V. are provided. Laundry, smoking (2 rooms), lunch and dinner buffets, bar, catering, and group rates of $60 for 10 or more rooms available. Visa/MasterCard accepted.

HOPE FARM (1780's), 147 Homochitto St. 445-4848. Open for tour all year round, Hope Farm has 4 rooms with a full Southern breakfast included in the price of $80. The twin-bed price is $70. There are easily accessible downstairs bedrooms for the handicapped and a free tour and access to the entire lavish garden property is included. Smoking is preferred to be done on the porch. The rooms come with private bath and children 6 and up are welcome. Reservations are usually necessary.

LINDEN (1785), 1 Linden Place off Melrose Ave., 445-5472. Open for tour all year from 10:00-3:00, this lovely home, which was seen in "Gone With The Wind", has 7 rooms for 2 for around $75. Free tour, Plantation Breakfast, private bath and handicap ramp come with this lovely property. Smoking is allowed and children 10 and over are welcome. Reservations aren't always needed.

MARK TWAIN INN Silver St. over the Under-The-Hill Saloon 446-8023. Open for tour all year from 11:00-2:00 with a water-level view of the Mighty Mississippi River. 3 rooms for 2, $55-$76, breakfast, private phone and T.V. available. Visa/MasterCard accepted.

MELROSE (1845), Melrose Ave. 446-9408. This wonderful mansion has 6 rooms for 2 at $100 which provides southern breakfast, private phone and bath. The rooms in the guest house have kitchen, T.V and laundry. There is also a nice fishing pond.

MONMOUTH PLANTATION (1818), John A. Quitman Parkway at Melrose Ave. 442-5852. Daily Tours 9:30-4:30 all year except Christmas. 14 spacious rooms, with breakfast buffet, tour, private T.V., phone and bath. Smoking is preferred on porch and children 14 and over are welcome and there is a Garden Cottage and elegant property with lovely pond, from $75-$150. Reservations are needed. All major Cards accepted.

MOUNT REPOSE (1824), Rt. 2 Box 159 Pine Ridge Rd. 446-7694. Open for tour all year, this fine country home has 4 rooms for 2 at $70, $75 with 2 of which are downstairs with easy access for the handicapped. Smoking is preferred outside. Comes with full southern breakfast, private bath and tour and has 2 ponds for fine bass fishing. Visa/MasterCard.

NATCHEZ PILGRIMAGE TOURS, Corner of State St. and Canal,Information and Tickets to most attractions, 446-6631.

PLEASANT HILL (1835), 310 S. Pearl St. 442-7674. Located within walking distance of all downtown businesses, this antebellum home has 4 antique-filled bedrooms with private bath, $75. There is a ground-level Garden Room for the handicapped. Open for tour during the Spring. Visa/MasterCard accepted.

PROPINQUITY (1790), Kaiser Lake Rd. off 61 N. 442-3292 or 442-1054. Open for Bed & Breakfast all year, this beautiful country home on the original Natchez Trace has 2 rooms for 2 with full Plantation Breakfast, private bath, fishing pond, and wine served in the evening. No infants or pets are preferred, but each room is filled with priceless family antiques and heirlooms as this house has been in the same family since 1832. The grounds are covered with camellias and azaleas planted by their ancestors.

RAVENNASIDE (1883), 601 S. Union St. 442-8015. Open all year, the 6 rooms for 2 are $65, with hot tub and pool table, private baths and refreshments are served from 5:00-6:00. Children are not preferred but there is access to the rest of the property.

SHIELD'S TOWNHOUSE (1860), 701 N. Union St. 442-7680. Open for tour during the Fall Pilgrimage. 2 rooms, 1 $80 and the suite $90. They provide freshly ironed linen daily. There are 2 1860 cottages with kitchen, private courtyard and parking garage. Deluxe continental breakfast, private phone, T.V., bath, and central garden with 15ft. fountain. Smoking is allowed and children over 12 welcome.

SILVER STREET INN (1840), "On The Mississippi", One Silver St. 442-4221. Open all year. 4 bedrooms, continental breakfast for 2 is $50. Smoking is allowed and there are private baths. T.V. is available in the parlor. All major Credit Cards.

STANTON HALL (1851), 401 High St. 442-6282, Reservations call 446-6631. Open all year are the 4 rooms for 2, which include plantation breakfast, and private bath, $75-$100. Reservations are needed. No pets but children 14 and over are welcome. Parties and receptions can be arranged. Visa/MasterCard/American Express.

TEXADA (1792), 222 S. Wall St. 445-4283. Open all year for tour, by appointment. There are the 4 rooms for 2 at $70, and there is a guest house for 4. No pets or children under 3, but smoking is allowed and there are private baths. Southern breakfast comes with this Territorial Legislature Hall, which was the first brick house built in downtown Natchez.

TWIN OAKS (1814), 71 Homochitto St. 442-6990. Open for tour all year, this lovely town home has 1 room for 2 at $75, which includes private bath, phone, and T.V. Breakfast is southern bacon and eggs, cold juice, home-baked banana bread and coffee. No small children or smoking preferred, but food is always in the refrigerator.

WEYMOUTH HALL (1855), One Cemetery Rd. 445-2304. Open for tour all year, this award-winning renovation has the 5 rooms for 2 at $60-$70. Includes large Plantation breakfast and private bath. Children over 13 are welcome with smoking outside. This is the only Greek Revival house that faces the river with its magnificent view. Visa/MasterCard.

Hotels & Motels

BEST WESTERN PRENTISS INN Hwy. 61 S. 442-1691, 1-800-528-1234. 128 rooms, meeting for 350, pool, restaurant with buffet, bar.

DAYS INN OF NATCHEZ 109 Hwy. 61 S. 445-8291, (800) 325-2525, 123 rooms restaurant, gift shop, handicap rooms available, tennis court, playground, Texaco Gas, pool. Visa, MasterCard, American Express, Carte Blanche and Diner's Club.

HOLIDAY INN OF NATCHEZ Hwy. 61 N. 442-3686, (800) HOLIDAY. Pool, playground, restaurant, bar, meeting for 60, buffet.

NATCHEZ EOLA HOTEL 110 N. Pearl St. 445-6000, meeting for 600, restaurant, buffet and bar.

NATCHEZ INN 218 John R. Junkin Dr. 442-0221, 38 rooms, pool, Visa and MasterCard.

RAMADA HILLTOP INN 130 John R. Junkin Dr. 446-6311, 1-800-272-6232, 1-800-228-3344 for groups & meetings up to 375, pool, restaurant, bar, gift shop, gazebo, 163 rooms. Beautiful view of the River.

SCOTTISH INN MOTEL Hwy. 61 S., 442-9141.

SHAMROCK MOTEL 700 Carter St., Vidalia, LA 336-4261

SHERATON NATCHEZ 645 S. Canal St. 446-6688, 148 rooms, 2 pools, bar, restaurant. Meeting rooms for 330.

THE SMALL MOTEL Airport Rd. off Hwy. 61 N. 446-7855.

TERRACE MOTEL Rt. 5 Box 38 Hwy. 61 N. 445-5516, 21 rooms, pool playground, Visa, MasterCard.

TRACE MOTEL 345 D'Evereux Dr. 442-7441.

Steamboats at Natchez Under The Hill. Courtesy of Prince & Natchez Photo Lab.

POINTS OF INTEREST

When it come to things to see, that is all that Natchez is. There are endless places to go and stories to hear. With the Antebellum structures that remain and the proliferation of stories that accompany them there are more sights and wonders than any one person could explore in a lifetime.

Most of the listings below can also be further explained in Chapter 1.

ANTEBELLUM HOMES

Before the War Between The States, Natchez afforded a great wealth of cotton planters who were called cotton kings. These men found some of the largest fortunes in the world here and built great mansions to symbolize this success. From all over the world, they had the finest of building materials and luxuries shipped in at phenomenal expense to adorn these delightful homes. When the Civil war came, it came to other places. Because Natchez had no military defence and little strategic value, the city and its structures were left intact. They still stand as they did in the nineteenth and even the eighteenth century. Though a couple of them were added onto by the original owners and other Pre-Civil War owners, some of the homes were always kept in perfect order and repair, while the rest have only undergone minor restorations. Very few, if any have been structurally altered since their lavish inception so long ago. A listing and description of the Per-Civil War homes and buildings is found in Chapter 2

"Laurel Hill Church" Courtesy of Historic Natchez Foundation.

BLUFF

The bluff offers a breathtaking view of the Mississippi River from downtown Natchez and used to be called the Promenade by the several European governments that occupied the territory. It was their parade ground for marching soldiers and a courting-ground for strolling Natchez lovers. Once watched over by Fort Rosalie and a lighthouse that was destroyed by a tornado, the two-hundred foot Loess Bluff commands a view of many miles north, south, and west over the river valley and the beautiful Louisiana countryside.

Geologists say that, during the ice age, a glacier created the flatlands below and winds blew silt and dirt and piled it into the hills now called Loess Bluff, which banks the east shore of much of the Mississippi River. Then the glacier melted and the Mississippi River itself was carved out of the soil. As time has gone by, the Mississippi has worked its way eastward into the Loess Bluff, eroding away the bank and causing the great hill to fall off in giant chunks and leaving the flat, silty, mineral rich land behind.

The Bandstand in the bluff park was named in honor of Katherine Grafton Miller, who first started the Pilgrimage to Natchez and opened the doorway for a vital tourist economy.

Houses on "The Bluff" Courtesy of Historic Natchez Foundation.

CONFEDERATE PAGEANT

Every year during, and since the beginning of, the Spring Pilgrimage the garden clubs have held the grand event. Like a play of time capsules in the form of tableaux, talented actors and dancers, adorned in the dress of the times, portray the changing episodes of history that make up the life of this community and the development of the United States Of America. No where else is the feeling of the old south so fully experienced as the players in each scene are truly the descendants of those men and women who pioneered and settled this area. This is what Natchez is all about.

For schedules and prices see Entertainment Chapter 7.

DUNCAN PARK

A fully equipped recreational facility, with pool, tennis courts, playgrounds, putt-putt golf, 9 hole golf course (18 on the way), pavilion, baseball fields and nature trails, Duncan Park was donated by the descendants of Dr. Stephen Duncan for public recreation. The antebellum home, Auburn, a National Historic Landmark, was his residence when he lived here before the Civil War. The park is open until 10:00 P.M. Auburn is always on tour.

EMERALD MOUND

This great Natchez Indian Mound covers eight acres and is the second largest Indian Mound in the state. Built by the forerunners of the Natchez Indians, it was used as a sacrificial temple until the Natchez moved to the Grand Village Of the Natchez Indians, south on St. Catherine Creek.

GRAND VILLAGE OF THE NATCHEZ INDIANS

Discovered officially by La Salle and Tonty in 1682, this National Historic Landmark, was the center of government for the thirty tribes of the Natchez Indians. Though it was the central area of worship, and most people today think of Indian villages as a small, concise arena of activity, the geography of the local land and the convenience of living off the land they afforded, this village was really spread out over a great expanse of land. There was no need for the group protection necessary, so the huts were spaced out an eighth to as much as a quarter of a mile apart. This association with the land induced a good and proud way of life. Jean Penicaut, the French carpenter, described the tribe and their village as polite and affable, but when they felt overrun by the Europeans, they retaliated and massacred nearly every man at Fort Rosalie. Penicaut was one of the few that was spared. The French laid siege to the Grand Village, then chased them across the river to Sicily Island where they were annihilated as a tribe forever. Some of the Natchez survivors who escaped and lived with other tribes have descendants that still survive. Located at 400 Jefferson Davis Blvd., Hours: 8:00 A.M.- 5:00 P.M., Monday-Saturday, 1:30 P.M. - 5:00 P.M., Sunday. Free. 446-6502.

JEFFERSON MILITARY COLLEGE

Built in 1802 as the first educational institution in the Mississippi Territory and named for President of the United States and Philosophical society, Thomas Jefferson, the college grounds housed the Methodist Church where the first Territorial Assembly was held in 1817 to form the government and Constitution of the State Of Mississippi, giving Washington the name "Birthplace of MS Statehood." Also held on the grounds was the treason trial of Aaron Burr. The famous painter, John J. Audubon taught there; President of the Confederacy Jefferson Davis was a student there as a young boy and Andrew Ellicot, surveyor of the 31st parallel, as well as Andrew Jackson and his troops camped there.

There are full facilities and a nature trail. The grounds are open all day and the buildings are on tour from 9:00 A.M. - 5:00 P.M. Monday - Saturday and 1:00 P.M. - 5:00 P.M. Sunday, Free. Located on Hwy. 61 N. in Washington, MS 442-2901.

Adams Light Infantry at Jefferson College. Courtesy Mississippi Dept. of Archives & History.

JERSEY SETTLEMENT

In 1772 Captain Amos Ogden received a 25,000 acre land grant for his efforts in the French and Indian War. Along with the two wealthy New Jersey planters named Richard and Samuel Swayze, he started this little community on Kingston Rd. near the Homochitto river. The church and the cemetery are still there.

LIBERTY PARK

Located at Liberty Rd. and Sgt. S. Prentiss Dr., the park includes a completely equipped, 2,000 capacity Convention Center, regulation baseball park, horse arena and stables, enclosed pavilions, fairgrounds and National Guard Armory.

MEMORIAL PARK

Located on the grounds once used as the cemetery of St. Mary's Cathedral on Main St., the City Park is a beautiful resting and relaxing spot. There is a fountain with goldfish and the Confederate Monument honoring the Adams County's Civil War dead.

NATCHEZ-ADAMS COUNTY PORT

To see how Natchez conducts trade with the mighty Mississippi River, call the Port Facility at 442-2561 to find out if they're loading up until 3:30 P.M. At the end of River Terminal Rd. near the industrial park.

NATCHEZ CITY CEMETERY

One of the prettiest old remaining cemeteries in the south, this beautiful, rolling landscape is a picture of grace, while the view of the river is also magnificent. The fascinating home, Weymouth Hall stands in awe on the edge of the cliffs, overlooking the river and the colorful cemetery. Located off the end of Linton Ave., 445-5051.

NATCHEZ STATE PARK

Offering picnicking, hiking, camping, biking trails, swimming, fishing, tennis, cabins, putt-putt golf and a 310 acre lake, this state funded park is located 10 miles north off Hwy. 61 at Stanton, 442-3208, 442-2658.

NATCHEZ TRACE PARKWAY

Charted by the Chickasaw Indians who roamed the country in their trading pursuits, the 450 mile Natchez Trace became a mail and trading route between Nashville and Natchez. It was also the way home for flatboatmen who floated down the Mississippi River with a load of goods to Natchez and New Orleans, where they sold everything, including the boat, and travelled back to the north in groups, often loosing their earnings and their lives to marauding bands of thieves and murderers. Today the Natchez Trace offers picnicking and nature trails and sites like the Emerald Mound. Open always as a National Park Highway it is a beautiful and serene drive.

NATCHEZ-UNDER-THE-HILL

Once noted for its sin and iniquity, the Natchez landing turned away and impressed many people travelling the river with the opinion that Natchez was a scum town when, in reality, the beautiful town above its high bluff was an absolute contrast.

As trade opened up on the Mississippi River, vice and immorality came into play at this den of violence as ships were hastily dismantled to erect buildings for the illicit trade which might bring quick fortunes. Many writings of visitors and tradesmen called the people who lived there the "scum of the earth," who conceived trap doors over the river and tunnels into the bluff for the propagation of their heinous crimes. There were few legitimate businesses and many brothels, where the shameless life was cheap and easy to find right on the streets and sidewalks of this malignant culture.

Though there were once many streets and even a horse race track in this hell hole, nature herself has seen to its demise through earthquakes, tornados, and the consistent chewing of the Mighty Mississippi River, which have sent over 160 acres of Natchez-Under-The-Hill into oblivion.

OLD MARYLAND SETTLEMENT

Colonel Wood of Maryland brought pioneers to the area off the Natchez Trace, Hwy. 553, Church Hill, where several homes and the beautiful Christ Church still stand.

OLD SOUTH WINERY

Natchez' own winery offers tours and testing daily, 508 Concord, 445-9924.

U.S. GOVERNMENT NATIONAL CEMETERY

The resting place of U.S. and Confederate Military Servicemen and Women who have died bravely over the course of many wars and battles in the Service of Freedom. This a solemn and reverent park located past the Natchez City Cemetery and open to the public.

THE NATURE OF NATCHEZ

Natchez is a nature wonderland because of its abundance of water, mild climate and rich soil. Everything, every variety, every species, every form of living thing seems to do well. I know this is a slight exaggeration, but there is a very large variety of natural life that thrives so well here.

Trees. Mississippi is one of the largest timber suppliers in the nation with more than half the state covered in forest. Though the magnolia is the states official flower, it is also the official tree which grows so abundantly. Other trees are the tung, gum, oaks (white, water, live, post, red, chestnut, blackjack, pin, shumard, black), many varieties of pine, cedar, locust, hickory, willow, cypress, cottonwood, tupelo, poplar, sycamore and maple.

Flowers. Plants can be seen flowering here at anytime during the year, even in the dead of winter if we have "false spring." Natchez is well known for its flowers, especially azaleas, of which there are many varieties of rare ones. Others include many types of roses, Virginia creeper, honeysuckle, buckeye, wood violets, goldenrod, purple vetch, black-eyed Susans, pink primrose, dwarf dandelions, aster, wisteria, cinnamon fern, magnolias, dogwood, redbud, mountain laurel, sumac, hazelnut and huckleberry.

Animals. Though some of the game have been hunted to near extinction, like the bison, bear, wildcats, beavers, muskrats, wolves, and cougars, they have started a comeback which allows all of them but the bison to be occasionally seen in the wild. The bison is preserved in captivity. Other animals that grow abundantly are the rabbits, deer, field mice, foxes, squirrels, raccoon, skunk, opossum. There are still mink, woodchucks, weasels and shrews.

Fish. Bass are found in several varieties as well as perch, bream, carp, buffalo, catfish, crappie, and pickerel.

Birds. The mockingbird is the official state.bird but many others are found and watched: cardinal, warblers, quail, hawk, eagle, woodpeckers, sparrow, ducks, geese, cormorants, dove, raven, blackbirds, buzzards, crane, orioles, thrushes, finches, teal, turkey and owls.

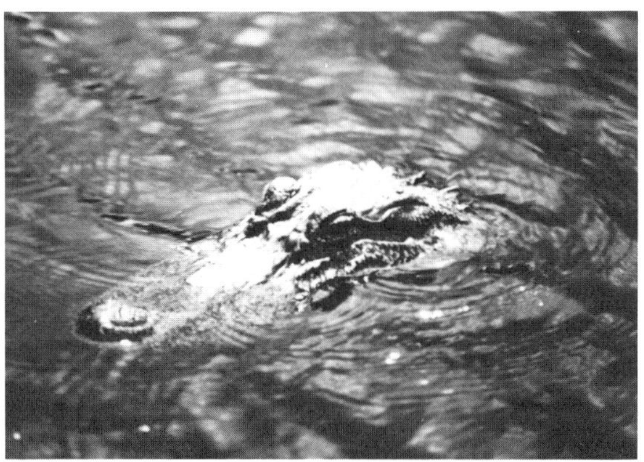

Alligator in the Louisiana Swamps just across the river.

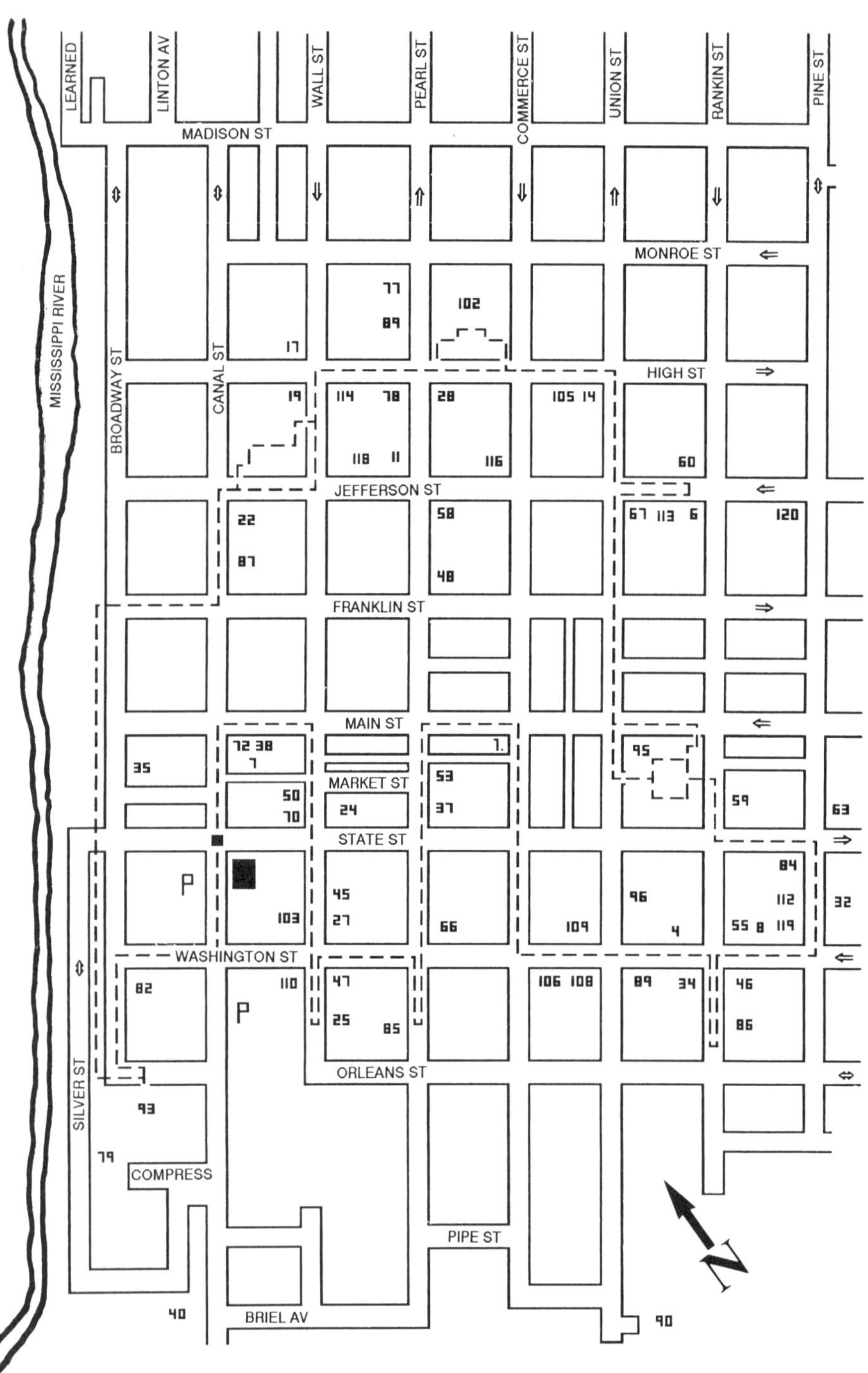

MISSISSIPPI RIVER
LEARNED
LINTON AV
WALL ST
PEARL ST
COMMERCE ST
UNION ST
RANKIN ST
PINE ST
MADISON ST
MONROE ST
HIGH ST
JEFFERSON ST
FRANKLIN ST
MAIN ST
MARKET ST
STATE ST
WASHINGTON ST
ORLEANS ST
PIPE ST
BRIEL AV
BROADWAY ST
CANAL ST
SILVER ST
COMPRESS
P
P
N
77
89
102
17
19
114
78
118
11
28
116
105 14
60
22
87
58
48
67 113 6
120
72 38
7
35
1
50
70
24
53
37
95
59
63
45
27
103
66
109
96
4
84
112
55 8 119
32
82
110
47
25
85
106 108
89
34
46
86
93
79
40
90

5

Walk-A-Bout-Natchez

Walking is a marvelous way to spend time, and walking through history is one of the many pleasures Natchez offers. If you like the constitutional, you'll love the walking tour of Natchez. Even if you would rather ride bicycle, it can be done in style and enjoyment on these old streets. You can even use this guide to study the homes and history when you take a carriage ride through this place in the past. Whatever you do, do it to learn about your heritage that is the gift of this great country and the people who pioneered it.

As you can see, the map on the facing page is self explanatory. The dashed line circles the entire downtown area, passing most of the interesting sites

Washington Street before. Courtesy of Bill Stewart.

Jefferson St. Methodists Episcopal Church at Union Street. Courtesy of Bill Stewart.

close enough to study their beauty. We thought it unnecessary to include arrows for walking direction because the walk may be made in either direction. Simply follow the line and reference the numbers indicated to the corresponding numbers in Chapter 2. Restaurants are all over the downtown area, so if you'd like to stop and eat lunch or take a short break, you'll be sure to find a spot. Where the line splits, it is an alternate route that allows studying the architecture more closely.

This walk is just under 3 miles in distance but can be made much shorter or longer as desired. No real planning is necessary except in the time you allow to yourself to spend. If you ride a bicycle or just want the exercise of fast walking without stopping, the trip can be made in an hour. But we suggest taking your time, studying the unique designs of the buildings, taking lots of pictures and generally trying to feel your way around. It is important to try and get a deep understanding, a thorough impression of what Natchez was in those days. When you approach a house or building that really catches your eye, walk all around it taking many pictures, study the details of the roof line, the window designs, doors and columns. Stare at it for a while and try to picture what it and the people who lived in it were like. There is a very gratifying feeling in *knowing* things that cannot be explained with words, and open-minded meticulous study can bring this gratification.

It is best to take the whole or at least half the day. Start in mid morning so you can stop to eat lunch about half-way through and buy more film.

Park at the Pilgrimage Tour Headquarters at one of the two P's shown on the map, Canal St. Depot at State St. Start walking in a counter-clockwise direction North on Canal St. If you have the time, stop in on the antique and gift shops to browse. If you see things you'd like to buy and don't want to carry them, you can have the proprietor mail them or drive by later and pick them up. You might even take a tour of some of the homes.

If you don't have much time or energy when you reach the Bluff overlooking the river, you can cut straight back up State St. to your car. If you do have extra energy, you might consider walking down to the bottom of Silver St. on the river bank. It's nearly a mile round trip, and it's quite a hill, but it's well worth it.

Until the opening of the Natchez National Historical Park, there is very little that's new in this town; that's the glory of it. The streets are the same, the buildings have been kept intact, many of the people were born of many generations of Natchezians, the pioneers of this colony, and all of this permanency stands in view of modern civilization for the visitor to enjoy. This is why it is important to indulge oneself into these histories, if for no other reason than to better understand yourself.

Main Street in Natchez as it was. Courtesy of Bill Stewart.

Award Winning Photo by T.G. McCary.

6

RECREATION

As in the days of cotton fields, when children rolled barrel hoops with a stick, adults as well as children need to recreate themselves in the simplicity of play. It it this revitalization that keeps one young at heart. Natchez still offers an abundance of activities with its diverse culture and economy. Many kinds of people live here and they do many different things to give their lives that extra something, that fulfillment of their spirits that is otherwise empty when neglected. It keeps this old town of Natchez young in spirit. It's a simple thing to do–play!

Since I consider R&R so important and its diversity so great, I have devoted much space to this chapter. There is a little something for everyone: from hobbies to sports, collecting to hiking, it's here. And while this is only a partial list of activities and where they can be found, it gives a sampling of the fun available and hopefully stirs the imagination into some activity that is good for the heart and soul.

Shopping and collecting are divided into different categories such as art, antiques, gifts and can be referenced in this section. If something is not listed, it may be found in the phone book or by contacting the Chamber of Commerce or Natchez Pilgrimage Tours, Chapter 3.

ANTEBELLUM HOMES

Natchez has over 500 of these pre-Civil War buildings, some dating back to the 1700's. Nine of the homes are National Historic Landmarks. Descriptions of the individual homes are found in Chapter 2. For tourist information see Chapter 4.

ANTIQUES

Natchez is an antique town and boasts one of the largest selections of antiques in the country. The Pilgrimage Garden Club has held their annual Antiques Forum here for the past fourteen years and the Natchez Garden Club holds their Annual Antique Show & Sale here. Listings of antique shops come under Shopping in this section. Also See Chapter 8 for dates of Antique Shows.

ART & CRAFTS

The famous painter John J. Audubon lived and taught in Natchez and many of his fine works as well as originals by other great artists remain in the homes and hearts of Natchez' citizens. This historic town has several nice Art Galleries featuring local artists and limited edition prints. For Arts & Crafts Shows see Events in Chapter 8. More information on art can be obtained from the Natchez Art Association, 310 N. Wall St., 446-5053. Shops are listed under Shopping below.

BALLOONING

Many people in Natchez own their own balloons. Balloons have been seen and photographed in Natchez since the turn of the century, but for the past three years in October (See Chapter 9), The Great Mississippi River Balloon Race and Festival has painted the skys over "Old Man River" with the nation's best balloons and pilots. They offer $5,000 in prizes, famous names concerts, arts & crafts, face painting, street dancing, juggling, fine food and drink, and thousands of people rocking the beautiful bluff behind Fort Rosalie overlooking the mighty Mississippi River. For information contact the Historic Natchez Foundation , Adams County Chamber Of Commerce or Natchez Pilgrimage Tours (For numbers See Chapter 3).

BICYCLING

Touring the countryside on bicycle is one of the best ways to see Natchez. Hundreds of people making up dozens of tour groups cycle through Natchez every year. Most of them come by way of the scenic Natchez Trace. Renting bikes is an increasingly popular form of recreation everywhere and Natchez is no exception.

Also, there are several annual biking events held here like the Bicycle Marathon, see Events Calendar in Chapter 8.

Natchez Bicycling Center 334 Main St. 446-7794.

BIRD WATCHING

Positioned at the "tip of the funnel" of the Mississippi flyway, Natchez is a superb sight to photograph and study birds. Literally millions of ducks, geese and other birds that come to the area are photographed and studied by many of the locals. Natchez has its own bird sanctuary at Natchez Under The Hill just below the old Mississippi River bridge. A pretty little sitting park, the sanctuary faces a panoramic view of the mile-wide river. The Natchez Trace and its nature trails is also a wonderful place to see birds and other nature as well as Duncan Park and The Homochitto National Forest.

BOATING

The Miss-Lou area abounds with lakes, streams and reservoirs. If you have your own boat or would like to rent one, contact one of the sporting goods dealers below for information. In the past at Natchez Under The Hill, small Paddle Wheelers have offered short cruises up and down the mighty Missis-

sippi River. This is a fascinating trip as it floats along the muddy banks draped in hanging willow trees, where Mark Twain picked up some of his inspiration. Many movies have been filmed on those same waters.
Lonnie's Canoe Rental 605 Hwy. 49 N., Seminary (601) 722-4301.
Tate S.J. Boat Camp Lake Concordia, Ferriday (318) 757-9939.

Boating in old Natchez. From the Prince Collection. Courtesy of Natchez Photo Lab.

BOWLING

Many bowling tournaments are held in the Miss-Lou area. The bowling alley in Natchez has an Arcade with pool tables, 16 lanes, open and league bowling, group rates to churches and schools, birthday parties, and it's easy to find. Driving about a mile from the Mississippi River bridges on John R. Junkin Dr. it will be on the left across from County Market, K-Mart and Western Sizzler.
Rivergate Bowling Lanes 285 John R. Junkin Dr. 442-8436.

CAMPING

For this recreation, the Nature Of Natchez speaks for itself. There is an abundance of everything.
Lee's Cash Grocery Fishing 7 days a week all year long. Hwy. 569, Ferriday. LA (318) 757-4449.
Natchez KOA Located on beautiful Lake Concordia off Hwy. 84 just outside Vidalia, LA 71373, only 5 minutes from Natchez, the KOA offers all facilities like hookups, drive-through paved sites, showers, recreation center, electricity, gasoline, propane, water slide, convenience store, boating rentals, fishing, hunting, and playground (318) 757-2397.
Natchez State Park Hwy 61 N. 446-6544. See Chapter 4.
Oak Harbor On Lake Concordia Complete camping services, convenience store, boat ramp, pier, showers. Ferriday, LA (318) 757-2397.

Plantation Mobile Home Park Paved and lit streets and lots, laundromat, T.V. cable, water and sewer hook-ups and swimming pool. 1/2 mile south on hwy. 61 route 61, 445-9894.
Spokane Resort Air conditioned rooms with kitchenettes, camping facilities with all hook-ups. Located 20 miles from Natchez on magnificent Lake St. John in Louisiana, one of the largest lakes in this area, it offers fishing, hunting, boating and boat rentals are available. (318) 757-4303.
Sportsman's Lodge Lake Concordia, Ferriday, LA (318) 757-4381.
Traceway Campgrounds Complete hookups, water, sewer, lights, hiking and biking trails, and swimming pool. Hwy. 61 N. near Natchez Trace Entrance (601) 442-3624.
Whispering Pines Travel Trailer Park Water, electricity, sewer hook-ups, recreational vehicles. Hwy. 61 N., 3/4 mile from Natchez (601) 442-3624.

CARRIAGE RIDE

Horse-drawn buggies available at State and Canal St. and the Eola can take you on a nostalgic trip through history and tree-lined streets. This is a most delightful experience in a city that is part of the past. 446-6631. See Recreation.

COLLECTING (Rocks, Artifacts, Bottles)

Mississippi is one of the largest suppliers of gravel and stone in the country, making it an excellent place to collect rocks. One of the best places to find Agates, Paint rocks, Geodes, and even precious gems is along the river banks at the Bird Sanctuary by Natchez Under The Hill. This is also a great place to find some of the rarest and valuable antique bottles and artifacts which get washed up by the river stages or when the water level drops to expose them. Another place to find bottles and artifacts is in numerous antique shops around the area (See Antiques). Collecting on the Natchez Trace and the Grand Village of the Natchez Indians is illegal. Find a list of Shops under Shopping below as well as Antique and Gift Shops. Art, Antique Shows and Flea Markets are found in Chapter 8.

FISHING

The Miss-Lou area offers more fishing spots than a person can find in an entire lifetime. The list of large bodies of waters is exhaustive not to mention the ponds and streams both commercial and natural that are filled with catfish, all forms of bass, bream, white perch, hybrid and others. There's Lake St. John, Lake Concordia, several Old river beds, the Blue Hole, The Mississippi River, Black River Lake, Red River Wildlife Management, Homochitto National Forest, Pipes Lake, Clear Springs, and commercial catfish ponds. There are also dozens of fishing tournaments throughout the year. Free Fishfinder Magazines are available at newsstands or check with one of the sporting goods dealers or the Mississippi Game and Fish Commission or the boat ramp attendant at either a chosen fishing hole or boat rental for information about what's biting. Professionally Guided fishing is available with all equipment and boat for two people for $150 all day, contact Dave

Blackman at Sportsman's Paradise below.

The law in Mississippi states that anyone 16-65 years of age and not disabled is required to carry a license on them for either hunting or fishing. An All Game Hunting & FIshing License for a Mississippi resident costs $13. For non-resident it's $70, but the prices drop considerably for a three-day ($25) or seven-day ($30) permit. Game Fishing is legal the year round. Licenses for Louisiana can be obtained at the Concordia Parish Sheriff Dept., Hwy. 84 Vidalia, LA (318) 757-3162. For Mississippi:

The Sports Center-Rex 107 N. Pine St. 442-4827, 442-7951 and The Natchez Mall 442-7906.

Sportsman's Paradise 1642 Carter St. Vidalia, 336-7122.

FLYING

Natchez offers some of the most beautiful scenic views of the Mississippi River Valley and countryside from hilltop but it is truly magnificent from the air. If you are not a pilot, check with one of the airports to see if planes and pilots are available.

Algin Flying Service Jonesville Airport, LA (318) 339-8722.

Natchez-Adams County Airport Airport Rd. off 61 N. 442-5171.

Silver Mount Corp. 603 Franklin St. 445-2150.

South Central Group Inc. Canal St. Depot 442-1603.

GIFT SHOPS

For Souvenirs, Antiques, fine gifts, collectibles, Arts & Crafts and gourmet foods see Shopping in this Chapter.

Duck Hunting in the Natchez Area. Photo by T.G. McCary.

GOLF

The public golf course at Duncan Park was donated to the city by the descendants of Dr. Stephen Duncan, who provided in the title of their home, Auburn, for the surrounding property be used only for public recreational facilities. It is a challenging 9 hole course (18 are in planning) that navigates ancient oak, tremendous pines and majestic sycamores. This beautiful course has fine greens and very reasonable fees ($5 for all day). Carts are available and walk-ons are welcome. Frequent tournaments are held and visitors are encouraged to participate. There is a complete pro shop, instructor, food and beverages, and locker room. 8:00 A.M.-8:00 P.M. every day year round. 442-5995.

Bellwood Country Club Cargill Rd. 442-5493. This is a private 18 hole Country Club. There is a swimming pool and tennis courts and complete facilities.

Panola Woods Country Club Clayton Hwy., Ferriday 757-2301 Beautiful 9 holes, private.

HEALTH CLUBS

Highland Health & Recreational Club 600 Highland Blvd. 445-9988. Nautill us, weights, steam, sauna, pool, racquetball, basketball nursery, juicebar and aerobics.

Natchez Racquetball Club Lower Woodville Rd. 442-5977.

HIKING

Natchez is covered with forests, hills, lakes and streams. Wildlife abound at every turn. It is interesting to walk along the Bluff or through downtown Natchez along the river bank. In Duncan Park there is a nature trail and the Natchez Trace has many fine trails. The Grand Village of the Natchez Indians makes a fine stroll. Even a walk through one of the old cemeteries is enjoyable. Wherever you are in Natchez, walking in any direction can prove interesting and educational. Historic Natchez Foundation has a detailed walking guide to downtown and there is our own Walk-A-Bout-Natchez in this guide.

HISTORY & RESEARCH

Those interested in history will find Natchez fascinating. Of all its benefits and curiosities, history is the forerunner of the list. Her buildings and homes are testimony to that. Visit them and your imagination will stretch with an undeniable longing for those times so long remembered and so short-lived.

Archaeologists Unlimited is mainly a Historic Restoration Consultant responsible for refurbishing many of the homes in this area. They also do cultural resources surveys 445-8464.

The Historic Natchez Foundation offers use of the many books and journals written in or about Natchez through the ages. Conducted as a small library, the Foundation will be more than glad to help you find interesting material that can be studied there. They also offer resources and consultation on historic restoration, 109 N. Commerce St. 442-2500.

Judge George W. Armstrong Library is the largest public library in this area. It contains over 70,000 books, as well as records, films, micro films, newspapers and has ample room for comfortable study. S. Commerce St., 445-8862. For hours see Chapter 3.

Concordia Parish Libraries in Vidalia and Ferriday have materials pertinent to history across the river. 405 Carter St., Vidalia, LA 336-5043 and 709 N. 3rd St, Ferriday, LA 336-8467, Clayton, Hwy 65, Ferriday 757-6460.

Catahoula Parish Libraries, Sicily Island, (318) 389-5804, Harrisonberg, LA 744-5271, Jonesville, 205 Pond Rd. 339-7070.

HUNTING

The Miss-Lou area offers some of the finest and most abundant hunting grounds and facilities available. There have even been deer in the heart of downtown Natchez. Legally hunted are deer, turkey, dove, duck, geese, squirrel, raccoon, bobwhite quail, gallinule, coot, woodcock, snipe, rail, wild boar, bobcat, fox, rabbit, crow, blackbirds, opossum, frogs and others have been hunted within a mile of downtown Natchez.

The protected wildlife are all birds of prey (eagles, hawks, osprey, owls, kites and vultures) and other nongame birds. Pheasant shall have no open season. Also protected are black bear, Florida Panther, gray bat, all sea turtles, gopher tortoise, sawback turtles (black-knobbed, ringed, yellow-blotched), black pine snake, eastern indigo snake, rainbow snake, and the southern hognose snake.

Homochitto National Forest is available to the public as well as the Red River Wildlife management area in Louisiana. Other public sections are also available. Information on guided expeditions, year round hunting and seasonal hunting schedules and licenses can be obtained from the Dept. Of Wildlife Conservation, P.O. Box 451, Jackson, MS 39205, 961-5300, or one of the sporting goods stores. (See Fishing above for License info)

PHOTOGRAPHY

Since 1851 when the brothers Marsh and Henry Gurney, professional photographers, came to Natchez and the father and son team, Henry C. and Earl Norman created superb works of art on film, Natchez has been as photogenic as a place could be. The Gurneys and the Normans continually kept abreast of the development of the times and purchased state-of-the-art equipment to ensure the quality of their work. Today is no different; new developments have progressed at a phenomenal rate and Natchez is evermore photogenic for both the professional and amateur photographer, as Natchez is displayed more and more in travel magazines and guides, homes, gardens, and architectural magazines, history books and newspaper articles. Natchez is a photographer's delight. Photo equipment, repairing, film, and one-hour developing can be found in the listings below.

LeBouef Studio & Camera Center 518 Main St., 445-4852.

Natchez Photolab Natchez Mall, John R. Junkin Dr., 445-3686.

Southern Exposure Camera Supply Store 487 John R. Junkin Dr., 442-8488.

PICNICKING

Duncan Park Duncan Ave., playgrounds, barbecue grills, tables, putt-putt and 9

hole golf courses, swimming pool, tennis courts and nature trail, (no camping).
Natchez State Park Fenwick Station Rd. 442-3208, 442-2658.
Natchez Trace Parkway abundant facilities, off Hwy. 61 N., emergency number at Ranger Station 445-4211.
The Bluff Park at the western end of Main St. overlooking the Mississippi River is a wonderful place to lay out a blanket and enjoy a magnificent view. There are also many outlying parks and lakes and campgrounds available for picnicking (See appropriate listing).

PUTT-PUTT GOLF

Duncan Park at the end of the drive on the left 442-5132.

RACQUETBALL

See Health Clubs above.

SIGHTSEEING

Natchez offers possibilities galore for the sightseer, whether it be tours, boating excursions or just walking through the old town. (See appropriate listing or Points Of Interest in Chapter 4.)

SHOPPING

Souvenirs, antiques, fine gifts, you name it, all the shops are listed below.
Antiques-J.E. Guercio 701 Franklin St. 442-5941.
Audubon Gallery 103 S. Broadway St. 442-2329.
As You Like It 410 N. Commerce St. 442-0933.
Auburn Gift Shop (Auburn) 400 Duncan at Auburn St. 442-5981.
BJ's Kwik Shoppe 47 Seargent S. Prentiss Dr. 445-2207.
Brown Barnett Dixon's Fine Gifts 511-515 Main St. 442-2931.
Brown & Church Ltd. 631 Franklin St. 446-6436.
Capricorn Gallery 403 Main St. 445-4235.
Connelly's Tavern Canal at Jefferson St. 442-2011.
The Corner Cupboard 197 Homochitto St. 442-8933.
The Country Bumpkin 508 Jefferson St. 442-5908.
Country Cottage 104 Belle Grove Dr., Vidalia, LA 336-5549.
Country Treasures 206 Washington St., 445-8577.
Creative Crafts & Gifts 804 Carter St., Vidalia, LA. 336-5357.
Crepe Myrtle Antiques 702 Franklin St. 442-1857.
DAR Gift Shop (Rosalie), Orleans at Broadway.
Days Inn of Natchez Hwy. 61 S. 445-8291.
Dianne's Frame Shop 408 Franklin St. 445-5444.
Fair Heavens Gift Shop 1053 N. Pine St. 442-9139.
Geri's Antiques 125 Commerce St. 442-1263.
Good Things Canal St. Depot 442-1626.
H. Hal Garner Antiques & Interiors 610 Franklin St. 445-8416.
Harper's Antiques And Interiors Canal at Jefferson 442-5217.

International Gifts & Novelty 308-C N. Pine St. 446-8917.
J.E. Guercio 701 Franklin St. 442-5941.
Lane's Pharmacy 1631 Carter St., Vidalia 336-5661.
Latham's Flower Shop Sedgefield Rd. 442-1545.
Lea's Valu-Rite Pharmacy Magnolia Mall, D'Evereux Dr. 442-6412.
Lessley's Pharmacy 115 Jefferson Davis Blvd. 446-6311.
Longwood Gift Shop 140 Lower Woodville Rd. 442-5193.
Lower Lodge Antiques 617 Franklin St. 442-2617.

"Magnolia Hall" Antebellum gift shop and museum. Courtesy of Natchez Convention Center.

Magnolia Hall Gift Shop 215 S. Pearl at Washington St. 442-7259.
Mama's Old Store Stanton St. 442-3847.
Medical Arts Pharmacy Tracetown Shopping Cntr. 442-3681.
Miss Dawn Antiques 205 State St. 442-0099.
Molasses Flats Antiques 220 Main St. 445-4591.
Mollies Follies 405 N. Franklin St. 445-2261.
Moreton's Flowerland 629 Franklin St. 442-4321.
Morgantown Pharmacy 114 Morgantown Rd. 442-2833.
Natchez Party Time 8 Southgate Shopping Cntr. 446-8533.
Personal Touch Gifts & Cards 114 Morgantown Rd. 442-2833.
Plantation Gourmet 116 S. Canal St. 446-7315.
Ramada Hilltop Gift Shop 130 John R. Junkin Dr. 446-6311 #298.
Reynold's Jewelry 1707 Carter St., Vidalia 336-5506.
Rushing Leather Craft & Screen Arts 201 Main St. 446-7667.
River Boat Gift Shop Natchez-Under-The-Hill 442-9711.
Sancha Gallery 716 Franklin St. 442-1052.
Sandpiper Country Antiques & Gifts Canal St. Depot 442-4742.

Southern Mississippi Antiques & Jewelry 620 Franklin St.442-5216.
Southern Seasons Inc. 118 N. Canal St. 445-8378.
Southern Traditions Canal St. Depot 445-5988.
Stanton Hall Gift Shop 401 High St. (Stanton Hall) 442-6282.
Tass House Antiques 111 N. Pearl St. 446-9917.
Trash & Treasures Providence Rd. 442-3274.
Treasures Unlimited Gifts 308 S. Wall St. 442-0432.
Uptown Arts 201 State St. 442-3533.
Van Court Antiques 205 State St. 442-4449.
Victory Antiques & Gifts 526 Franklin St. 442-5527.
Yesteryears 118 Canal St., 446-8393.
You Name It Natchez Mall 445-9588.

SIGHTSEEING

Under Points Of Interest in Chapter 4 there are many things to explore in the Natchez area. On the facing page is a picture of the Old Sunken Natchez Trace.

SKATING & SKATEBOARDING

Hide-A-Way Skating Rink party arrangements, 504 Fisk Ave. 445-8965.

SKIING

If you have your own boat, Lake St. John and Lake Concordia are both very large, prime skiing lakes from early spring to late fall. There is usually a very small boat-ramp fee. For directions contact place of lodging or area map.

SWIMMING

Natchez offers many natural water areas for swimming as well as several pools. Lake St. John and Lake Concordia are good swimming holes. Though not recommended, even the river has been used. Duncan Park, Connelly's Tavern and Bellwood Country Club, Highland Health & Racquet Club have pools.

TENNIS

Duncan Park 8 lighted courts, 442-1589.
Bellwood Country Club 6 courts, 442-5493.

WALKING

Walking in downtown Natchez is best done with Walking Guide in hand. Walk-A-Bout-Natchez in Chapter 5 makes one great sweep of the downtown area, covering most of the interesting subjects. Another good, detailed walking guide is available through The Historic Natchez Foundation, 109 N. Commerce St. 442-2500.

WINERIES

Old South Winery daily tours, 508 Concord St. 445-9924.

Old Sunken Natchez Trace dug down by hoof, foot, wagon wheels and time.
Courtesy of The National Park Service.

Bennie Boone and Weeta Colebank play the "Mississippi Medicine Show"
Courtesy of the Natchez Convention Center.

7

ENTERTAINMENT

Since before Manuel Gayoso de Lemos, Governor of the Louisiana territory and commandant of the Natchez Settlement (see Chapter 1), brought in the charming Spanish attitude of entertained living, Natchez has loved "getting together" in an assortment of diversions. In the earliest days there were elegant parties given in the fun-loving Spanish style. In the antebellum times, when cotton was booming and Natchez was one of the richest planting areas in the world, magnificent Soirees were thrown constantly by the southern ladies in their bell-shaped hoop skirts and the proud gentlemen of the south. It seems that the great mansions were built around this time for just that purpose.

When the Civil War came, the gaiety was interrupted for a time while 1,444 men left the Natchez area for battle. It was a time of confusion and heartbreak because at least 510 of them died in that vain cause. Federal troops occupied Natchez and the abundance of Confederate money was worthless. But Natchez was not to break under the pressure. Only seven years after the war the Natchez economy began to pick up where it left off with a new revival of true southern living. The younger ones, who remembered those elegant balls, had grown old enough to initiate their own life of leisure in the form of picnics. Quite often were these elaborate picnics given in honor of life itself and adorned with all the grace that compliments it.

New customs and games were invented while old ones were revived. There were horse races, steamboat races, elegant dining, folklore and country music, carriage rides through the streets and countryside, and men lazily congregating around storefronts to chew the fat. In time, all manner of sports and entertainment were enjoyed by both ladies and gentlemen: balloon riding, floating circuses. And the finest goods were imported from all over the world to wear and decorate their homes.

Today, the tradition continues with parties (less elaborate but just as fun) and Sunday "get-togethers", where large families and friends collect and eat and talk of the past, the times, and the future, sports, politics, and hunting. Mardi Gras is celebrated with parades and balls. Every year during the Spring

Pilgrimage, elaborate balls are held for the Kings and Queens of the Garden Clubs with the thought going back through time to those who, so long ago, made Natchez what it is today. It is a time honored tradition in Natchez, especially for the many families who have been here for more than 200 years, to enjoy life to its fullest. Join us in living life the way it was meant to be lived.

Night Life

The alphabetical listings in this section are exclusively nightlife, indoor activities. Daytime activities are listed in the recreation section. There are also sections on points of interest (Chapter 4) and events (Chapter 8).

CONFEDERATE PAGEANT

We bring this under a separate heading because it deserves special attention. It is similar to theatre performance, yet it means much more to those who attend. It is a "must see" if you are in or can get to Natchez when it is shown, so we give it top billing in our entertainment section.

Depicting ever so truly the glamour and beauty of the historical periods of the Natchez area, the Confederate Pageant, sponsored by the Natchez and Pilgrimage Garden Clubs during Spring Pilgrimage, is recognized world-wide as one of the best Pilgrimage performances. Quite elegant and magical, each tableaux takes you back into a period of historical importance to Natchez and this great nation. The dance and music performed by various trained groups–some as dancers and others as singers and actors–all dressed in the attire of the period. Most of the elegant ball dresses in the "Soiree" are designed by dance instructor, choreographer and designer, Allie Minette Middleton. Some of the gowns cost very large sums of money. For more than 30 years she has worked with seamstresses and Garden Club members to thrill the audience in what has become more popular and brought more tourists to Natchez every year since 1932, when the Pilgrimage was first organized by Katherine Grafton Miller and friends.

This is a very special event that is unique and endearing. The old south is best experienced in this way as it will leave one with the wonderful feeling that they have gone back in time and been a part of their heritage that is this great nation. A true look into history cannot be described by words but can be seen and felt and lived here in what is part of a living history.

The two-hour performance starts at 8:30 P.M. on Monday, Wednesday, Friday, and Saturday during the Natchez Spring Pilgrimage (March 11 - April 9, '89) at the City Auditorium, 207 Jefferson at Canal. Admission is $7.00 and reservations are needed because it is always sold out! Obtain tickets from the Natchez Pilgrimage Tour headquarters: Canal at State St. or P.O. Box 347, Natchez, MS 39121, (601) 446-6631, 1-800-647-6742. Future dates can be found in Chapter 4.

DINING & DANCING

In the not too distant past there was a trend away from the high-life, and it wasn't until 1964 that the Legislature allowed liquor to be sold in Mississippi but only through state-licensed package liquor stores (10:00 A.M. - 10:00 P.M.) and lounges. No liquor is sold through package stores on Sunday, Christmas or Election days while the poles are open. The liquor laws in Mississippi are strictly enforced, however Natchez is an official resort area and certain exceptions are allowed.

RESTAURANTS

Below is a selection of preferred restaurants with indications of what they specialize in and the services they offer, including any handicap or special amenities such as live entertainment or dancing. Though the list includes most of the better fast-food and sit-down restaurants, Natchez is a resort area subject to trends in the trade wherein many good establishments go in and out of business. It is always best, during Pilgrimage, to call and even make reservations before going.

THE ANNEX TEA ROOM 209 Franklin St. 446-6544, at the foot of Ellicot's Hill, "The Gathering Place" has exotic salads, soups, sandwiches, desserts, homemade bread and other light meals. It's moderately priced and can arrange catering or banquets for bridge or luncheons any time. Open 9:30-3:30 Mon.-Fri. and Saturday during Pilgrimage. Visa/MasterCard.

BELLEMONT SHAKE SHOP Hwy. 61 S. 445-4581. Open 9:00-6:30 Mon.-Sat. Serves good, inexpensive sandwiches and hot meals for 5 minute take-out or patio dining. Lunch specials during the week.

BEST WESTERN PRENTISS INN Hwy. 61 S. 442-1691. Open 5:30-10:00 P.M. Buffet Mon.-Fri. for breakfast ($3.95) and lunch ($4.95, $6.95 Sunday) and supper during busy times. Full Service menu for up to 200, banquets, catering. Chef's specials every night. Music, darts and T.V in the lounge and live entertainment. All major Credit Cards.

BONANZA STEAK HOUSE At the main entrance of the Natchez Mall. Open 11:00-9:00 Sun.-Thrs. and 11:00-10:00 Fri.-Sat. $4.69 dinner buffet every night and 11:00-2:30 lunch buffet for $3.99 as well as daily specials. Full service menu, banquet arrangements. Bus driver eats free and 10% off of paid group check. Thanksgiving dinner. Visa/MasterCard.

BURGER KING Trace Town Shpg. Ctr. 442-6711. Carter St., Vidalia, 336-5522.

CAKES BY LINDA 145 Homochitto, 446-7868. Mon.-Sat. 9:00-5:30 Tues.-Sat. Not a sit down restaurant but good take-home food. "Special Cakes For Special Occasions" Specializing in New Orleans Style Po-Boys on fresh baked breads, good creole hamburgers, cakes, cookies, and pastries. Catering for desserts and sandwiches. Take-out service in 10 minutes.

CANTINA 207 State St. 446-6456. Open 11:30-2:00 Mon.-Sat. and 5:00-9:00 Tues.-Sat. Their 6 tables and small bar are kept busy through the end of the week but no reservations are necessary to get the good Tex-Mex food (Mexican with a Texan flavor). Specials on drinks sometimes. They aren't large enough for a bus load but catering can be arranged. Visa/MasterCard/American Express.

CARRIAGE HOUSE RESTAURANT 401 High St. 445-5151. Open daily for lunch at 11:00 -2:30 and dinner 5:30-8:30 during Pilgrimage at Stanton Hall. Fine Old Plantation Dishes, Cajun and dinner theater. Special arrangements for tours, receptions and parties. Handicap parking, daily specials, banquet and catering arrangements, group rates from Natchez Pilgrimage Tours. In the Gay 90's building behind Stanton Hall. "Nationally famous delicious southern cooking." Visa/MasterCard/ American Express.

COCK OF THE WALK Natchez-Under-The-Hill 446-8920. Fine catfish on the banks of the Mississippi River in the Old Cypress Shanty (1832). Catfish, hushpuppies, fried dill pickles and mustard greens. Mon.-Sat. 5:30-9:30, Fri.-Sat. 5:00-10:00. Open 7 days during Pilgrimage. $7.50-$8.95 for the "Finest Catfish on the MS River." Group rates for bus load and catering for more than 285. Live piano bar whenever open. Holiday events. Visa/MasterCard/American Express.

DAIRY QUEEN 245 John R. Junkin Dr. 442-3200.

DAYS INN "DAYBREAK RESTAURANT" 109 Hwy. 61 S. 445-8291. Open 24 hours for good, inexpensive full service menu. Daily lunch specials Mon.-Fri. and bus drivers eat free. Music and fast take-out. All major Credit Cards.

DOMINO'S PIZZA 421 Hwy. 61 S. 445-6006

DOUG'S RESTAURANT 410 Main St. 446-7193. Open Mon.-Fri. at 5:30-2:00 and 6:00-9:00 and Sat. and Sun. 6:30-2:00. They are nearly always busy because the home cooking is good and inexpensive, but no reservations are needed. The daily lunch buffet is $4-$6. No credit cards.

EOLA RESTAURANT 110 N. Pearl at Main St. 445-6000. Juleps has breakfast buffet from 7:00-10:00. New Orleans lunch specials and Po-Boys, 11:00-2:00. Dinner of Seafood Gumbo, Blackened Catfish, Shrimp Creole, Chicken Bon Femme, Tournadoes Eola and other specials at Cafe La Salle, 5:00-10:00 every day. Sunday brunch, 11:00-2:00. All Major Credit Cards.

THE FARE 109 N. Pearl St. 442-5299. Open Mon.-Sat. at 11:00-3:00. Good specialty sandwiches, fresh garden salad, imported coffees, and homemade dessert at reasonable prices. Visa/MasterCard.

FISH FRY 1100 Carter, Vidalia 336-7229. Open 11:00 AM daily until late. Seafood, Po-Boys, Gumbo, and good catfish. Carry out available. Banquets and catering available and bus driver eats free. Visa/MasterCard.

HARDEE'S 339 D'Evereux Dr. 442-1053.

HOLIDAY INN "RAINTREE RESTAURANT & LOUNGE" Hwy. 61 N. 442-3686. Raintree restaurant open every day 6:00 AM-10:00 PM. Daily buffet and dinner specials.

HUNAN CHINESE RESTAURANT Hwy. 61 S. 442-2121. Open every day for lunch and dinner. Chinese take-out orders and lounge. Specializing in Hunan & Schechaun style dishes. 11:30-2:30 and 4:30-10:00 Mon.-Fri. Fri.-Sat. open until 11:00. Sun. at 12:00-10:00. Major Credit Cards.

JOHNNY'S PIZZA Hwy. 61 S. 442-0037 and 1701 Carter St, Vidalia, 336-9611

KENTUCKY FRIED CHICKEN Hwy. 61 N. 445-5781. Hwy. 61 S., 442-0992.

MALT SHOP 4 Homochitto St. 445-4843.

MAMMY'S CUPBOARD Hwy. 61 S. 442-9554. Home cooking, southern meals, Hot Tamales, Catfish, great hamburgers, and hot roast beef. This 1940 building shaped like a giant Mammy is a classic southern delight with good food and country atmosphere. Open until 9:00 PM daily and Sunday, 10:00-3:00. Daily Specials on their Hot luncheon Menu. Visa/MasterCard.

MCDONALD'S Trace Town Shpg. Ctr. 445-8493 and 1600 Carter St., Vidalia 336-9063

MOM'S is open 24 hours at 1639 Carter St., Vidalia 336-5642.

THE NATCHEZ DINER Hwy. 61 S. 445-8291. Open all day every day. Breakfast, shrimp, chicken, steaks and sandwiches.

THE NATCHEZ LANDING LTD. 11 Silver St. at Natchez-Under-The-Hill, 442-6639. Tues.-Sun. 6:00-10:00. Combination platters, Fried and B-B-Q Shrimp, catfish, excellent ribs, chicken, steaks, and sandwiches. On the river bank in renovated antebellum building. Visa/MasterCard/American Express.

NOTHIN' FANCEE DELICATESSEN 112 N. Commerce St. 442-6886. Mon.-Fri. at 8:00-5:00 and Sat. 11:00-4:00. New York style sandwiches at good prices. Daily specials and take-out in 10 minutes. Home of the MS Catfish Burger, the Mouffalatta, the Greek Gyros, Diet Plates and Po-Boys.

THE PARLOR RESTAURANT 116 S. Canal St. 446-8511. Open 10:00-10:00 Mon.-Sat. Homemade soups, chili, burritos, nachos, fajitas, desserts and good steaks. with moderate prices fast service, good drinks and daily specials. Catering handled by Anna Ernst. Visa/MasterCard accepted.

PIZZA HUT N. 227 D'Evereux Dr. 442-2221 and Hwy. 61 S., 446-8421, and Concordia Shopping Center., Vidalia, 336-4334.

PIZZA HUT DELIVERY, 172 Sgt. Prentiss Dr. 445-9700.

PO-BOY HUT 610 Carter, Vidalia, 336-8330, 336-9944. Cajun B-B-Q and over 40 different sandwiches on fresh-baked break, onion rings, catfish and ice cream. Take out available.

POPEYE'S FRIED CHICKEN D'Evereux Dr. 445-9482. & Carter St., Vidalia, 336-5269.

THE PRENTISS CLUB 211 N. Pearl at Jefferson St. 445-9480. Lunch Mon.-Fri. 11:00-2:30, Diner Mon.-Sat. 5:30-until. Fine Dining. Closed Sunday Evenings.

RAMADA HILLTOP INN 130 John R. Junkin Dr. 446-6311. Open every day 6:00-10:00. Southern buffet for breakfast and lunch Mon.-Fri. Gourmet buffet on Sunday. Special seafood selection Friday evening. The best prime rib in the south on Wednesday night. Wide selection: Cajun, Italian, great Southern Fried Chicken, Steaks, Sandwiches, Seafood. Lounge with live entertainment Tuesday-Saturday. High on the bluff with panoramic view of the Mississippi River. Handicap facilities and moderate prices. All major Credit Cards.

SANDBAR RESTAURANT 106 Carter St. Vidalia 336-5173. 11:00-10:00 daily. All sorts of good seafood and catfish selections in a casual friendly atmosphere. Located at the food of the bridge in Vidalia. Major Credit Cards.

SCROOGE'S OLD ENGLISH PUB & RESTAURANT 315 Main St. 446-9922. Open 11:00-11:00 Mon.-Sat. Full service menu and bar. Trifles, salletts, fish, fowl and beef. Daily luncheon specials, dinner and children's specials. Nice unique atmosphere with collection of early Norman Photographs. Visa/MasterCard/American Express.

SHANTYMAN'S COVE Hwy. 61 N. at St. Catherine Creek 442-5957. Wide selection of steaks, chicken, seafood, B-B-Q, fried vegetables, chicken livers, taco salad, chef's salad. Bus tours welcome. Open Mon.-Sat. 5:00-9:30.

SHERATON NATCHEZ 645 S. Canal St. 446-6688. Tues.-Sat. 6:00-10:00. Lamb, Veal, Quail, Pheasant. Lounge with entertainment.
SHONEY'S RESTAURANT 26 Sgt. S. Prentiss Dr. 442-3761
SONIC DRIVE IN 294 Sgt. S. Prentiss Dr. 446-8351.
SOUTH CHINA RESTAURANT 495 John R. Junkin Dr. 442-8548. Open from lunch until. Specializing in Szechuan & Cantonese selections. Lunch specials every day. Carry out and catering.
SOUTHERN KITCHEN Hwy. 61 N. 446-7164. Mon.-Sat. 6:00-9:00. Home cooking, seafood buffet. One mile outside city limits.
TACO CASA Natchez Mall 446-6046. Mon.-Sat. 11:00-9:00, Sun. 12:30 -6:00. Variety of mexican dishes. Located at the main entrance of the Natchez Mall.
TAMALE HOUSE 12 4th St. 442-4492.
UNDER-THE-HILL "OLD FRONT WATERING HOLE" 33 Silver St. at Natchez Under The Hill 446-8023.
WEST BANK EATERY Levee Rd., Vidalia 336-9669. 11:00 AM until. Fine Creole, steaks and special seafood dishes.
WESTERN SIZZLIN STEAK HOUSE 288 John R. Junkin Dr. 446-7052. Open 11:00-9:30 Sun.-Thrs. and 11:00-10:00 Fri.-Sat. 5 luncheon specials like children under 12 eat free from children's menu, one for each paid adult. Bus driver and tour director eat free. Banquet and catering. Handicap parking and ramp. Visa/MasterCard.

"Shield's Townhouse"

LOUNGES

Below is a list of prime party areas, a few of which serve food, but mostly those that serve up a good time with music and dancing, or lounging comfort in soft, plush elegance, or the historic pub-like atmosphere of the Natchez Trace days, when travel was by horse or foot and the need for rest and relaxation was great. Lounges and restaurants can also be found at most major hotels.

THE CORNER S. Canal St. 442-2546.
DIMPLES 324 Main St. 445-9464.
NUTT'S FOLLY PUB 140 Lower Woodville Rd. at Longwood 442-5193.
PEARL STREET CELLAR 211 N. Pearl St. 446-5022.
SCROOGE'S 315 Main St. 446-9922.
THE SPORTS PAGE 279 John R. Junkin Dr. 445-8814.
UNDER THE HILL SALOON 33 Silver St. 446-8023.
Y-VONNE'S LOUNGE Jefferson at Rankin St. 446-8845.

MOVIES

Natchez Mall Cinema IV Natchez Mall, 442-5218.
Tracetown Twin Cinema Tracetown Shopping Center, 442-5432.

THEATRE

City Of Natchez Theatre 207 Jefferson St. 442-9164.
Carriage House Restaurant 401 High St. at Stanton Hall 445-5151.
"Moonlight & Magnolias" is a dinner theatre presentation depicting the elegance of the past. It plays during the Fall Pilgrimage.
"Voices Of Hope" Amos Polk's Spiritual Singers perform this inspirational program during the Fall Pilgrimage. Also includes a traditional plantation dinner. Sunday, Tuesday, Thursday at 7:00 P.M. $15.00 admission. Reservations are necessary.
Natchez Little Theatre Playhouse 319 Linton Ave., P.O. Box 1232, Natchez, MS 39121, 442-223. The Natchez Little Theatre puts on famous shows all year round but is renowned for its "Southern Exposure" done annually.
"Southern Exposure" is a satire on the Pilgrimage first produced in Dallas at the Margo Jones Theatre. It played on Broadway in New York and on television, starring Elizabeth Montgomery. The play is presented every year during Spring Pilgrimage (March - April) by the Mayweather Hall Players on Tuesday, Thursday, Friday, Saturday and Sunday at 8:30 PM. Admission: $7.00. Reservations are necessary. See Natchez Little Theatre.
"Mississippi Medicine Show" is a two-hour river boat musical variety production characterizing an evening on the Mississippi river. It plays during the Fall Pilgrimage Monday, Wednesday and Saturday at 8:30 P.M. and available year round for groups of 25 or more.
"The Drunkard" written during the 1800's to discourage drinking, this comical play brings the lead characters from desperation to enlightenment. Performed by the Gaslight Players at the Sheraton Ballroom, Monday, Wednesday and Saturday at 8:00 P.M. Tickets available through Pilgrimage Tour Headquarters, Canal at State St.

The Great Mississippi River Balloon Race & Festivsl. Photo by Andrew Fisher.

8

EVENTS CALENDAR

Over the past years Natchez has grown in popularity among conventioneers, associations, tours and groups who hold big meeting events annually. Consequently, more and more annual and seasonal events are held at the hotels and the Convention Center. Most of the events which take place at the hotels must be found through the individual hotels or groups. We have included those held in other places like the parks and the Natchez Mall, so a general idea of what's going on during the season can be discerned.

Many of these dates are tentative, and where there is no date given it is an event that usually takes place annually and has not been formally planned. It is always best to check with the Natchez Convention Center & Visitors Bureau, (800) 647-6724.

– January –

LOUISIANA BUSINESS COLLEGE GRADUATION (19) at the Natchez Convention Center.

MARTIN LUTHER KING PARADE (16) Downtown Natchez.

NATCHEZ INDIANS AFTERNOON OF STORYTELLING (28) at Grand Village Of The Natchez Indians, 446-6502.

AMERICA'S COVER MISS PAGEANT (28) at the Natchez Mall.

MARDI GRAS CHILDREN'S PARADE PRESENTATION OF MARDI GRAS ROYALTY (30, tentative) at the NATCHEZ MALL.

– February –

"HARVEY" (2-4) comedy production of the Natchez Little Theatre, 442-2233.

MARDI GRAS PARADE (3) in Downtown Natchez at 3:30, contact Melba Wells, 446-7187.

AFTER PARADE PARTY (3) At Ramada Hilltop (private), 446-7187.

MARDI GRAS BALL KREWE OF PHOENIX (4) at the Natchez Convention Center. Contact Melba Wells, 446-7187.

CALL-OUT MARDI GRAS BALL (4) 6:30 at the Natchez City Auditorium (invitation only.)

PHOTO'S FOR VALENTINE'S DAY (4) at the Natchez Mall.
NATIONAL WILD TURKEY FEDERATION (11) at the Natchez Convention Center.
KIRK FISHER MAGIC SHOW (11) at the Natchez Mall.
NATCHEZ CHILDREN'S HOME BENEFIT SHOW (18) at the Natchez Convention Center.
DOBSON PETTING SHOW (14-19) at the Natchez Mall.
FARMERS/AGRICULTURE SHOW (18) at the Natchez Mall.
RIVER CITY RUN & 10K RUN (25) In Downtown Natchez, sponsored by Bluff City Spiders, contact Sara Garcia, 446-6090.

"Choctaw" for shows on Art.

– March –

ANTEBELLUM HOMES ON TOUR (1) throughout Natchez, 446-6631.
SPRING CRAFT SHOW (4) at the Natchez Mall.
SPRING PILGRIMAGE (March 11- April 9), tours of 36 Antebellum homes, parties and entertainment, CONFEDERATE PAGEANT, SOUTHERN EXPOSURE.
CONFEDERATE PAGEANT (11- April 8) performances at the City Auditorium, 8:30 P.M., $8, 446-6631.
SOUTHERN EXPOSURE (14-April 9) Natchez Little Theatre production, 446-6631.
EASTER BUNNY ARRIVES (11) at the Natchez Mall.
ST. PATRICK'S DAY TABLOID (17) at the Natchez Mall.

SPRING FASHION SHOW (18) at the Natchez Mall.
CHAMBER OF COMMERCE DRAW DOWN & DANCE (18) at the Natchez Convention Center, 445-4611.
TRUE BIBLEWAY CHURCH (24-26) ELDER ANDRE RAMSEY, at the Natchez Convention Center.

—April—

HOME GARDEN SHOW (1) at the Natchez Mall.
KELLY MILLER CIRCUS (5) at the Natchez Mall.
SPRING CHARITY BAZAAR (8) at the Natchez Mall.
SOUTHERN REFLECTIONS ARTS & CRAFTS (8-9) at the Convention Center.

Ralph Macchio and Joe Seneca in "Crossroads." © 1985 Columbia Pictures Industries, Inc. Movie making in Natchez has become more frequent than an annual event. The Natchez Film Commission helps Hollywood companies find locations throughout the year. Photo by Ron Phillips. Courtesy of Columbia Pictures. All Rights Reserved.

SPRING PILGRIMAGE (March 11-April 9).
MAY DAY FESTIVAL ARTS & CRAFTS (30-May 1, tentative) at "Choctaw", Wall at State St., 446-5053.
JEFFERSON COLLEGE STORYTELLING (held in '88)
STEVE GIBSON CARTOONIST & CARICATURIST (14-15) at the Natchez Mall.
SPRING FLEA MARKET (15-16) at Liberty Park and Convention Center (contact Natchez-Adams County Home Extension Office for info, 445-8202.)
ANNUAL HEALTH FAIR (22) at the Natchez Mall.
WHEELCHAIR TENNIS TOURNAMENT (22-23) at Duncan Park.
SOUTH NATCHEZ SCHOOL CHORAL FESTIVAL (25) at the Convention Center.
PILGRIMAGE GARDEN CLUB EXHIBITS (26-28) at the Natchez Mall.

– May –

NATCHEZ GARDEN CLUB ANNUAL ANTIQUE SHOW AND SALE (5-7), largest show in the southeast.
YOUTH FISHING RODEO (6).
NATCHEZ MALL 10th ANNIVERSARY CELEBRATION (6).
ACCS GRADUATION (12) at the Convention Center.
50 MILE RUN (13) Natchez Trace Parkway.
FARROW AMUSEMENT (15-21) at the Natchez Mall.
METRIC MILE RUN (20) at Natchez Trace Parkway.
NATCHEZ MIDDLE SCHOOL AWARDS NIGHT (23) at the Convention Center.
9th ANNUAL CHAMPIONSHIP TRI-STATE RODEO (26-28) at Liberty Park.
NATIONAL TRIPLE SOFTBALL SERIES (26-29) at Liberty Ball Park.
LITTLE THEATRE PRODUCTIONS
3-DAY RACE (27-30).
TRINITY EPISCOPAL SCHOOL MAYFAIR
HOSPITALITY DAY in conjunction with HOSPITALITY WEEK at the Convention Center and Welcome Center.
BLUFF CITY PROMENADERS

– June –

ANTIQUE CAR SHOW (3).
NATCHEZ STEAMBOAT JUBILEE DAY Mississippi Queen-Delta Queen steamboat race, Floozie Contest, concerts.
UGLY TIE CONTEST (17) at the Natchez Mall.
STATE YOUTH BOWLING TOURNAMENT
NATCHEZ HIGH SCHOOL RODEO
JEFFERSON COLLEGE 3rd ANNUAL PIONEER WEEK for Children Ages 8-12 (19-23 or 26-20).

– July –

4TH OF JULY ANNUAL FIREWORKS (1) at the Natchez Mall.

GRAND VILLAGE OF THE NATCHEZ INDIANS DISCOVERY WEEK (10).
50's & 60's DAY-CAR SHOW-ELVIS (15) at the Natchez Mall.
SUMMER CRAFT SHOW (22) at the Natchez Mall.
GOLD WING RIDERS (29) at the Natchez Mall.
FOURTH OF JULY RODEO CELEBRATION, Tri-state Championship Rodeo with all proceeds going to Natchez Children's Home (contact Mike Ash, 446-8688.)

—August—

FASHION SHOW (5) at the Natchez Mall.
NATCHEZ CITY TENNIS CHAMPIONSHIP (5-6) at Duncan Park.
YOUTH CITY TENNIS TOURNAMENT (12-13) at Duncan Park.
FLINTKNAPPING SEMINAR ('88) at Grand Village Of The Natchez In-

"D'Evereux" on tour during Pilgrimage. Courtesy of Natchez Convention Center.

dian, 9:00 AM-5:00 PM, observe and participate in making tools and weapons from stone, 446-6502.

DRB, INC. ARTS AND CRAFTS SHOW ('88) at the Convention Center, 371-3829.

–September–

FALL FESTIVAL (Sept.-Nov.) many arts and crafts and entertainment events in the lovely spring-like weather.

HUNTING & FISHING EXHIBITS (2) at the Natchez Mall.

GYPSY (8,9,10,11 of '88) is a musical play about the life of Gypsy Rose Lee at 8:00 PM or 2:00 PM on 11th, Natchez Little Theatre, 442-2233.

INDIAN FESTIVAL (9-10 from 9:00 AM- 4:00 PM and 11 from 12:00 Noon-4:00 PM) Indian dancing, sports, crafts at Grand Village of the Natchez Indians, 446-6502.

CONFEDERATE OIL GOLF TOURNAMENT (15-18 of '88) at Bellwood Country Club.

HUNDRED MILE BICYCLE RIDE (18 of '88) 8:00 AM, 100 miles up the Natchez Trace. Registration fee: $13 before Sept. 10, $15 up to one week in advance, includes a sag wagon with fruit, juices, cookies, a patch, lunch, and a T-Shirt. Claudia Soper, 334 Main St., Natchez, MS 39120, 446-7794.

OLD NATCHEZ TERRITORIAL FAIR (18-23) 10:00 AM-12:00 AM at Liberty Park, Rides, livestock shows and exhibits, food.

SOUTHERN HOSPITALITY HOOPLA (22-24) at the Convention Center.

ANTIQUE SHOW (23) at the Natchez Mall.

COPPER MAGNOLIA FESTIVAL (23-24) 9:00 AM-4:00 PM at Jefferson College, regional crafts demonstration and sales, 442-2901.

SANCTIONED 100 MILE BICYCLE RIDE (24) at Natchez Trace Parkway.

CATHEDRAL GOLF & TENNIS TOURNAMENT (30-October 1) 9:00 AM at Duncan Park, 445-5682.

SAVOR OF THE SOUTH BIKE TREK (24-25) by the MS LUNG ASSN. and the Natchez Jaycees 8:00 AM based at the Prentiss Hotel with a two loop ride (40 miles) finishing at Longwood, second will start and end at the Prentiss. (30-100 mile ride). Tom Kearns, 362-5453.

ACCENTS ON ART ('88) Antebellum tours, tea and art exhibit at Choctaw.

GHOST STORYTELLING ('88) around the campfire at Jefferson College 6:00.

MAGNOLIA STORYTELLING FESTIVAL (30 of '88) at Connelly's Tavern, 7:00 PM.

–October–

GOLF & TENNIS TOURNAMENT (1) at Duncan Park.

NATCHEZ TRACE FRONTIER FESTIVAL (6-8) 8:00 - 5:00 at the Convention Center.

FALL PILGRIMAGE (7-28), tour of 32 Antebellum homes, entertainment.

COUNTRY DAYS ARTS & CRAFTS FAIR (8-9 of '88) at Convention Center, 10:00 AM-5:00 PM.

VOICES OF HOPE (9,11,13,16,18,20,23,25,27 of '88) 7:00 PM at Carriage

House Restaurant.

CRUSADE FOR CHRIST (9-14) at the Convention Center. Rev. W.C. Mazique.

MISSISSIPPI MEDICINE SHOW (10,12,14,17,19,21,22,24,26,28,29 of '88) at Natchez Little Theatre, 8:00 PM.

THE DRUNKARD (8,10,12,15,17,19,22,24,26,29 of '88) at Sheraton Grand Ballroom.

CATHEDRAL SCHOOL FALL FESTIVAL (14-15), Jean Van Uden, 445-8844.

FALL FLEA MARKET (14-15) at Liberty Park Fairgrounds.

OLD SOUTHWEST RENDEZVOUS (14-16) buckskinners at Jefferson College, 9:00 AM-6:00 PM, 442-2901.

CRAFT DEMONSTRATIONS at Grand Village Of The Natchez Indians

PIONEER DAY (19) 8:30 AM-12:30 PM at Jefferson College.

GREAT MISSISSIPPI RIVER BALLOON RACE National competition, concerts, arts and crafts, B-B-Q, camp out.

NATCHEZ GUN SHOW (22-23 of '88) 9:00 AM-5:00 PM at Convention Center.

GHOST TALES AROUND THE CAMPFIRE (27 or 28) 6:00 PM at Jefferson College.

MAGNOLIA STORYTELLING FESTIVAL (27-28), Berry Bateman, 442-9407.

"Twin Oaks" Courtesy of Natchez Pilgrimage Tours.

FALL ARTS & CRAFTS SHOW (28) at the Natchez Mall.
QUILT SHOW (29 of '88) 9:00 AM-5:00 PM at Jefferson College.
HALLOWEEN GHOST STORYTELLING (31 of '88) Jefferson College, 6:00 PM until.

—November—

NATCHEZ ON DISPLAY (4) at the Convention Center.
MISSISSIPPI INDIAN CULTURE Indian Pow Wow at Liberty Park Fair Grounds.
NATCHEZ YULETIDE TREASURES ('88) at Natchez Convention Center, craftsmen from over the U.S. demonstrate and sell their art.
NATCHEZ COUNTRY CHRISTMAS (10-11) at the Natchez Convention Center.
HOLIDAY FASHION SHOW (11) at the Natchez Mall.
SOUTHERN REFLECTION ARTS & CRAFT SHOW (19-20 of '88) at Natchez Convention Center (contact Sandra Steinhauer, Rt. 4, Box 90 HS, Falson, LA 70437, (504) 796-5853).
CHRISTMAS CRAFTERS (25-31) at the Natchez Mall.
16TH ANTIQUE FORUM, contact Pam Harris, 442-7234.

—December—

CHRISTMAS CRAFTERS (1-24), BELL CHOIR, DANCE PROGRAMS at the Natchez Mall.
INTERNATIONAL PAPER COMPANY CHRISTMAS DECORATIONS (1-25) at the Natchez Mall.
CHRISTMAS IN NATCHEZ (10-24), parades, tours of decorated antebellum homes, entertainment, exhibits, Renaissance Feast.
CHRISTMAS PARADE
NATCHEZ JUNIOR AUXILIARY CHARITY BALL at Natchez Convention Center.
CHRISTMAS PARADE downtown at 3:00 PM.
COCHON-DE-LAIT (pig roast) (10) at 12:00 PM in the Memorial Park on Main St.
CHRISTMAS CAROLS (10 of '88) at the river and downtown at 5:00 PM.
CHILDREN'S VICTORIAN CHRISTMAS OPEN HOUSE (10) at Jefferson College, children's church choirs and decorations.
VICTORIAN LUNCH ('88) at Magnolia Hall.
CHRISTMAS ON THE HILL at Ramada Hilltop, entertainment, dinner, fun.
HIGH TEA (14, 21, 28 of '88) at Natchez Eola Hotel.
HIGH TEA (16, 23, 30 of '88) at Annex Room on Franklin St.
TOURS OF ANTEBELLUM HOMES (10-24) at 9:00 AM-5:00 PM.
CARRIAGE TOURS OF THE HISTORIC DISTRICT (10-24) at 9:00 -5:00.
OLD SOUTH WINERY daily tours and tasting.
BRUNCH (10-24 of '88) at Natchez Eola, Sheraton, and Ramada Inn Hilltop.
NEW YEAR'S EVE (31) 10:00 AM-6:00 PM at the Natchez Mall.

III

FUTURE NATCHEZ

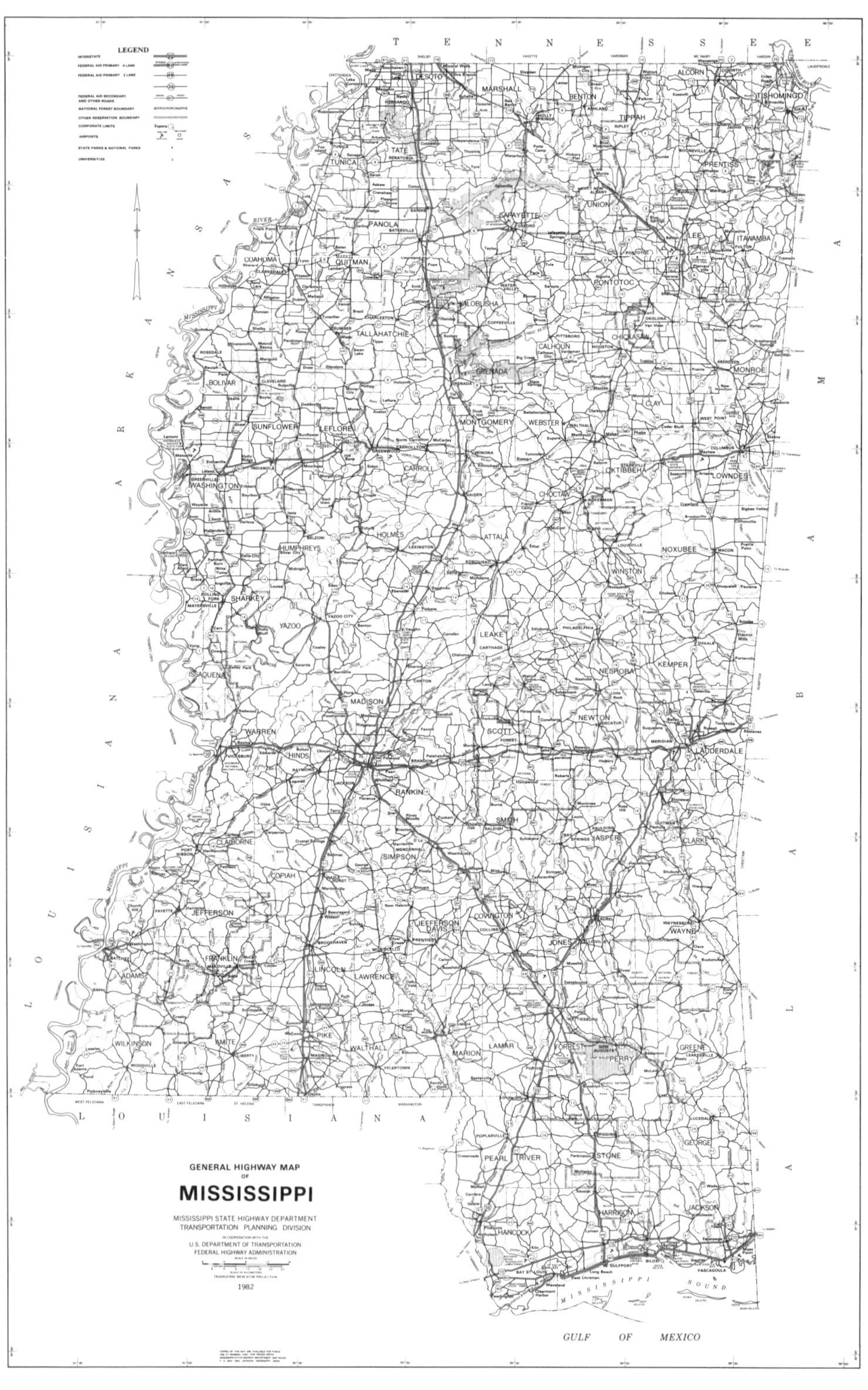

LEGEND
INTERSTATE
FEDERAL AID PRIMARY 4 LANE
FEDERAL AID PRIMARY 2 LANE
FEDERAL AID SECONDARY AND OTHER ROADS
NATIONAL FOREST BOUNDARY
OTHER RESERVATION BOUNDARY
CORPORATE LIMITS
AIRPORTS
STATE PARKS & NATIONAL PARKS
UNIVERSITIES
TENNESSEE
ARKANSAS
LOUISIANA
ALABAMA
GULF OF MEXICO
MISSISSIPPI SOUND
ALCORN
TISHOMINGO
DESOTO
MARSHALL
BENTON
TIPPAH
PRENTISS
TUNICA
TATE
UNION
LAFAYETTE
PANOLA
LEE
ITAWAMBA
COAHOMA
QUITMAN
PONTOTOC
YALOBUSHA
TALLAHATCHIE
CALHOUN
CHICKASAW
GRENADA
MONROE
BOLIVAR
CLAY
SUNFLOWER
LEFLORE
MONTGOMERY
WEBSTER
CARROLL
OKTIBBEHA
LOWNDES
WASHINGTON
CHOCTAW
HOLMES
ATTALA
NOXUBEE
HUMPHREYS
WINSTON
SHARKEY
YAZOO
LEAKE
NESHOBA
KEMPER
ISSAQUENA
MADISON
NEWTON
WARREN
SCOTT
LAUDERDALE
HINDS
RANKIN
SMITH
JASPER
CLARKE
CLAIBORNE
COPIAH
SIMPSON
JEFFERSON
COVINGTON
JEFFERSON DAVIS
JONES
WAYNE
FRANKLIN
ADAMS
LINCOLN
LAWRENCE
WILKINSON
AMITE
PIKE
WALTHALL
MARION
LAMAR
FORREST
PERRY
GREENE
PEARL RIVER
STONE
GEORGE
HARRISON
JACKSON
HANCOCK
GENERAL HIGHWAY MAP
OF
MISSISSIPPI
MISSISSIPPI STATE HIGHWAY DEPARTMENT
TRANSPORTATION PLANNING DIVISION
IN COOPERATION WITH THE
U.S. DEPARTMENT OF TRANSPORTATION
FEDERAL HIGHWAY ADMINISTRATION
SCALE IN MILES
SCALE IN KILOMETERS
TRANSVERSE MERCATOR PROJECTION
1982

9

THE FUTURE OF NATCHEZ & THE STATE OF MISSISSIPPI

The future of Natchez depends largely on the future of Mississippi, so we thought it appropriate to close this work with information that relates our fair city with this great state.

Mississippi is like the other great states of this country in that it is naturally beautiful. With its large variety of plants, animals, trees, flowers, and geographic regions, plenty of water and clean air, Mississippi has remained one of the prettiest states as it was in the beginning. The Indians, which have been here for at least 11,000 years, always lived in harmony with the land. Compared to the rest of the industrial nations of this world, civilized man has had little time to alter its natural beauty. Also, the general economy, which is agricultural and mineral, lends itself to this natural harmony, to conserve its original perfection in creation and in way of life.

Mississippi is a well balanced combination of elements, bestowing a mild climate and healthy atmosphere. She is 47,716 square miles, 296 of which is natural water (like its many rivers and streams and "old rivers"), 7,560 in swamps and river beds. There is also quite a large area of man-made flood-control area. Averaging about 400 feet above sea level, the highest point being 806 feet, the state is very consistent in climate, averaging around 65° (45° in winter, 80° in summer), with an average annual rainfall of 40 to 65 inches. The temperature rarely crosses the zero or the 100° marks. Yet, the seasonal changes are quite pronounced by bright colorful springs, lush green warm summers, glorious cool autumns, and bare haunting winters.

Mississippi is a Sports and Recreation Heaven, consisting of five general areas, each with its own different attractions.

The Hills area in northeast Mississippi offers a picturesque masterpiece of rolling landscape similar to that of Scotland. With its lakes, streams and trails, this is an outdoorsman's delight, providing endless sites for camping, hiking, canoeing, sailing, swimming, hunting, fishing, skiing and exploring historic sites like the home of William Faulkner, the birthplace of Elvis Presley, the ancient Indian mounds and dozens of museums. In this region are two national forests, seven state parks, four reservoirs and countless camping grounds. DeSoto County Welcome Center, I-55, P.O. Box 102, Hernando, MS 38632, (601) 368-9969.

The Delta area to the northwest is a broad expanse of color and beauty as far as the eye can see. Here is the very foundation of agriculture in the south, with its roots deep in the tradition of brotherhood and emotion that inspired the Delta Blues. Here in this breezy, flat land one will find many festivals celebrating life, museums preserving memories, music, hunting and fishing, recreational and sports facilities, four state parks, the Delta National Forest, campgrounds, and bicycling routes. Washington County Welcome Center, Hwy. 82 at Reed Rd., P.O. Box 6022, Greenville, MS (601) 332-2378.

The Plains area, east central, is known for its Indian mounds, where the Choctaw Indian Nation was born. Also a wonderful place for hunting and fishing and festival music with its scenic landscape and rocky streams. You will find harness racing, museums, country cooking and festivals, three national forests, one national and six state parks, four water parks, campgrounds, reservoirs, summer camps and conservation lakes. Canoeing, river fishing and bicycling abound. Lauderdale County Welcome Center, I-20, P.O. Drawer 30, Toomsuba, MS 39364, (601) 632-1142.

The Gulf Coast area, down south on the Gulf of Mexico provides white sand beeches, roaring surf and warm sun, deep sea fishing, scuba diving, wind surfing, sailing, marine parks and an abundance of festivals. There are music and dancing and sporting events as well as the NASA National Space Technology Laboratory. Antebellum homes and museums, golfing, tennis, canoeing, bicycling, two national forests, nine water parks, campgrounds and seven Wildlife Conservation Lakes. Hancock County Welcome Center, Waveland, MS (601) 533-5554. Jackson County Welcome Center, Moss Point, MS (601) 475-3384. Pearl River County Welcome Center, Nicholson, MS, (601) 798-8184.

The Heartland covering the central and southwest portions of Mississippi opens up varieties of doorways to the past, present and future. Where the steamboat paddle wheelers float lazily along Old Man River and the Antebellum Mansions pose in awe of the days when cotton was king; where parades of cars and floats and people of all colors and nationalities promenade by in celebration of the present Christmas Spirit of peace and understanding; where symbolic balloons fly high, University studies are conducted, arts and crafts are preserved, the environment is checked, safer forms of power and manufacturing and transportation are produced, and people work together to insure a better future for the generations to come.

The Battlement at Vicksburg stands as a monument to an era gone by. Needed by the Union army for its strategic importance, many brave soldiers from both sides gave their lives to control what is now one of the finest and largest Civil War museums in the country. Every battle line and battery are well preserved to remind this great nation of its fight for the freedom of all people.

Our Capitol city of Jackson is as progressive and upbeat as any city in the U.S., while presently preserving the heritage and history of its state. Theatre, symphony, art, opera and the International Ballet Competition are all part of this culture which strives to make Mississippi wealthier in spirit. Mississippi is a leader in moral issues and was first to pass laws for women's rights (Ch. 1).

Communities like Natchez, with it past and varieties of culture still in tact, with its better than average economy, striving for improvement, and its people, black, white, red and yellow, are learning to work side by side in peace and harmony with each other and the land, so their children will have a happy and bright future.

Wherever you go in Mississippi, you'll find something to suit your spirit and one thing common to all of this state, that wonderful thing called Southern Hospitality.

IN THE VICINITY

While you are in Natchez, it is good to know that other very different cultures are accessible with a short drive from here. It is no problem to drive to any one or even two of these attractions, explore many of their sites and return to Natchez before experiencing some of its well-known entertainment.

Vicksburg, about 75 miles up Hwy. 61. The National Military Park was the turning point of the Civil War, when Gen. Ulysses Grant surrounded the Confederate stronghold on the bluff for a 47 day siege. This is a site to see the entire battleground, trenches, cannons, cemetery and museums. When Vicksburg fell, the Mississippi River belonged to the Union and it meant the end of "the cause" as the south knew it and the beginning to the unity of this great nation.

Other sites and antebellum homes in Vicksburg are Anchuca, the Balfour House, Cedar Grove, Christ Episcopal Church, The Corners, Firehouse Gallery, Floweree House, The Galleries, Holy Trinity Church, the McNutt

Antebellum Mt. Carmel Church near Natchez. Courtesy of the Natchez Convention Center.

MILEAGE FROM NATCHEZ TO:

ABERDEEN	265
ALEXANDRIA, LA	65
AMORY	274
BATON ROUGE, LA	89
BAY ST. LOUIS	201
BILOXI	205
BIRMINGHAM, AL	346
BOONEVILLE	300
BROOKHAVEN	62
CANTON	127
CLARKSDALE	216
CLEVELAND	180
CLINTON	92
COLUMBIA	110
COLUMBUS	249
CORINTH	320
DALLAS, TX	400
GREENVILLE	154
GREENWOOD	172
GRENADA	205
GULFPORT	199
HATTIESBURG	139
HOLLY SPRINGS	286
INDIANOLA	168
JACKSON	102
KNOXVILLE, TN	605
KOSCIUSKO	172
LAUREL	147
LELAND	153
LITTLE ROCK, AR	265
LONG BEACH	194
LOUISVILLE	197
McCOMB	68
MEMPHIS, TN	295
MERIDIAN	192
MOBILE, AL	240
MONROE, LA	80
MONTGOMERY, AL	348
MOSS POINT	230
NASHVILLE, TN	487
NEW ORLEANS, LA	130
NEW ALBANY	283
OCEAN SPRINGS	208
OXFORD	255
PANAMA CITY, FL	350
PASCAGOULA	226
PENSACOLA, FL	285
PETAL	142
PHILADELPHIA	182
PICAYUNE	177
SHREVEPORT, LA	181
SOUTHHAVEN	284
STARKVILLE	226
TUPELO	272
VICKSBURG	75
VIDALIA, LA	1
WEST POINT	244
WINONA	182
YAZOO CITY	117

House, McRaven, the Old Courthouse Museum, Pemberton House, Planters Hall, Sargent S. Prentiss Building, St. Francis Xavier Convent, and William A. Lake House. At **Raymond**: Immaculate Conception Catholic Church, Jackson House, Magnolia Vale, Peyton House, Raymond Courthouse, St. Mark's Episcopal Church, and Mississippi College, Clinton.

Port Gibson, about 40 miles up Hwy. 61 N. or the Natchez Trace, developed out of a Spanish land grant to Samuel Gibson. It now contains a beautiful array of antebellum homes and buildings such as the Disharoon House, Englesing, Gage Home, Idlewild, McGregor, the Methodist Church, Planter's Hotel, the Presbyterian Church, St. James Episcopal Church, St. Joseph's Catholic Church, the Spencer Home, Temple Gemiluth, Van Dorn House, the Ruins of Windsor, Wintergreen Cemetery, and Grand Gulf State Military Park.

Woodville and **St. Francisville** area abounds with Antebellum history. Woodville, around 35 miles down Hwy. 61 S., has Argue Home, Bramlette Law Office, Bowling Green Plantation, Catchings House, Day Home, Elsinore, the Ferguson Home, the Foster Home, Hampton Hall, the Lewis Home, the Magruder-Scott Home, the Methodist Church, the Railroad Office, Rosemont, St. Paul's Episcopal Church, Woodville Baptist Church, the Woodville Public Library, the Woodville Republican, Cold Spring, and Desert Plantation. In St. Francisville, LA, another 20 miles or so from Woodville are places like Afton Villa, Asphodel Plantation, Audubon Memorial State Park, Catalpa, The Cottage, The Myrtles, Oakley House and Rosedown.

THE FUTURE

Natchez and Adams County are pushing ahead to make the area a better place to live for all its people. Many new proposals are in the works which will bring more people, more jobs, better economy and a cleaner state to live in. The Industrial Foundation and Chamber of Commerce has recently received visitors from Korea, Taiwan, Belgium and Japan with prospects of broadening their lumber and steel industries. Also several companies have either started new operations or expanded with the help of he Industrial Foundation.

WATERFOWL REFUGE

This is only one of the developments in the proposal stage to use 15,000 acres for a Wildlife Refuge for the conservation of Mississippi's nature.

UP WITH NATCHEZ

The Retail Services Committee has provided seminars to help the local businesses stay on their feet in these hard times.

NEW SAY

Our state and local Governments are allowing people to express their opinion more. The Chamber of Commerce co-sponsored a forum to give the public a chance to meet with candidates before the election.

OTHER DEVELOPMENTS

Thanks to our representatives we now have the Homochitto and Mississippi River Bridges along with new Highway developments and better prospects for future expansion.

The Mississippi River Bridges connecting Natchez to Louisiana. Photo by Andy Fisher.

NATCHEZ NATIONAL HISTORICAL PARK

Since the 1930's, when the Pilgrimage first began, a terminus for the Natchez Trace Parkway has been wanted and talked about. Years later, in the cities future plans were drawn proposed routes of the Trace Parkway and positive steps were taken to see it through. It never came true. It meant a lot to this little community that was suffering economically like the rest of the country.

The subject was brought up several times thereafter in hopes that enough interest would be generated to get things rolling. Still, Natchez was passed over for Federal highway funds on many occasions. When Natchez was provided with improvement money, and her highways began to see the development she so richly deserved, came the blessing of being recognized.

In 1987 the Mississippi Department of Archives and History was able to prove that Natchez played a very important role in this nation's history. In 1988 local and state organizations began laying plans to study the proposal's finalization and the Adams County Chamber of Commerce created a committee to promote the park.

The National Park Committee sent a twelve-man team here for study of the proposal and our Congressmen and Senators later introduced bills to authorize the park. After so many years of patient waiting, the bills passed both houses and President Reagan signed his O.K. to fund the planning stage of the Natchez National Historical Park.

According to officials, the park will mean a tenfold increase in tourist traffic–the equivalent of having a Fortune 500 company invest here.

The park will take up most of the downtown bluff area and some of the land along the river north and south. Plans aren't definite, but they will provide for a modern visitor's center and complex, one of the large Antebellum homes will serve as the terminus building, and the land along the bluff will be developed into a recreational park for everyone to enjoy. Many other ideas, like museums and theatres on the history of the area, development of the old Fort Rosalie site originally built in 1716, further preservation of Natchez Under The Hill, will be brought up. It will be a good addition to the heritage of this country and the education of its young.

Teaching our heirs about their history and heritage is not a physical thing, it is one which transcends the surface appearance. Education breeds understanding, this in turn leads to the development of better conditions for the human being, which eventually evolves into peace and happiness. With these philosophical and moral improvements will come the physical progress in technology, transportation, communication, industry, ecology and economy. All of this will make life better, not only in the physical sense but in understanding–the spiritual sense. The speed and safety of communication and transportation will bring our communities and the countries of the world closer together. It will also bring human beings closer together, that we may work together and live together in peace and harmony and happiness. Because of this understanding we will have taught ourselves.

It is because we humans are able to put forth the great amount of time and effort and financing it takes to accomplish a community development, like the

Natchez National Historical Park, that makes us better in this way. If the whole world got together to develop every corner of the earth, there would be peace.

As for this little area in and around Adams County, Mississippi, we look forward to this National Park and future developments of its kind. This whole area stands to gain many advantages from the park: jobs, greatly improved tourism, and better national and international promotion; and it will help her citizens to be even more proud of the community they live in – Natchez: "The City In History."

Near Natchez on location at the Ruins of Windsor, the film crew of "Crossroads" with director Walter Hill. © 1985 Columbia Pictures Industries, Inc. Photo by Ron Phillips. Courtesy of Columbia Pictures. All Rights Reserved.

INDEX

—A—

Aberdeen 29
abolitionists 26
Academy 25,53,60
African 54
Agricultural Bank 34
Airlie 34
Alabama 22,25
Allen, Betty 61
Allen, Major John L. 61
Americans 19,21-23
Antebellum Creed 28
Arkansas 14,25
Arlington 34
Armstrong House 34
Armstrong Library 92
Auburn 34,72
Audubon, John J. 29,41,35,45,78,88,94,120
Aunt Hill 35

—B—

Baker, Otis T. 30
Banker's House 35,39,70
Baptist 21,22,60,120
Barnes House 35
Bay St. Louis 15,17
Best Western 75,101
Bibles 21
Bienville 14-17,19
Biggs House 36
Biloxi 14-16,38
bit 20
Black List 19
Black Code 16,17
bluff 47,54,62,63,65,77,80,88,92,94,103,108,110,122
Bourbon County 20
Brandon Hall 36
Briars 36
Burling, Walter 24
Burn 36,54,72

—C—

Cadillac, Antoine de la Mothe 16
Canada 14,17,22
Canadians 16
capital 16,17,25
Cat Island 14
Catholic 22,53,120
Cato West 23
Chamber Of Commerce 57,58,62,87,88,109,122
Cherokee 37,70
Cherry Grove 37
Chester, Peter 19
Chickasaw 17,20,80
Choctaw 24,37,80,110,112,118
Chopart 18
Christ Church 20,21.48,80
Church Hill 36,43,48,80
circuses 29
City Hall 61
Civil War 30,61,76,78,79,87,99,118
Claiborne, W.C.C. 22,23,42,35
Clark, Daniel 20,50
clergyman 21,29
Cliffs 37,79
climate 24,117
Cloud, Adam 21
Colbert, James Logan 20
Cold Water Man 27
Coles Creek 19,43
Collel 20
colony 16,17,19,48
comets 25
Concord 20,37,63,80,96
Congress 19,22,46,54,
Connelly's Tavern 38,70,94,112
Continental Congress 19
Cottage Gardens 38
cottonseed oil 24
council of war 17
Council of Gentlemen 22
Court House 38,61
Crozat, Anthony 16
Curtis, Reverend Richard Jr. 21

—D—

D'Evereux 38
Dale, Sam 23
Davion's Rock 15
De Soto 14,15
Delta 110,118
Democrats 26,28
dime novels 29
Dixie 38
duel 27
Dunbar, Sir William 19-21,24,25,29,39,41
Dunleith 38,50,51,70,72

—E—

E.P. Fourniquet 34,35
Edgewood 39,70

INDEX

Elgin 39,40,70,72
Ellicot, Andrew 21,22,38,78,101
Elms 39,70
Elms Court 39
Elward 39
Emerald Mound 78,80
English Turn 15
entertainment 77,99-101,103,104,108,112,114,119
Eola Hotel 41,72,75,90,102,114
epidemic 41
Episcopalian
Essex 30,45
Evans-Bontura 39
explorers 14

—F—

faction 28
Fair Oaks 39,70
Father Davion 15
Fire Eater 28
First Church Of Christ Scientist 39
First Presbyterian Church 39,45
Fleur-De-Lys 16
floating stores 20
Florida 14,19,93
folklore 29
Foote, Henry Stuart 28
Forest House 39,41
Fort Adams 14,22
Fort Rosalie 15-18,22,40,50,77,78,88,122
Fort Pitt 19
Fort St. Peter 18
Fort Maurepas 15
fossils 25
France 14-17,19,23
Free Papers 26
French and Indian War 19,79

—G—

Galvez, Bernardo de 20
Gardens 40
Gayoso, Manuel de 20-22,42,36,37,65
Georgia 19,20,23,31
German Coast 17
gins 20
Glenburnie 40
Glenfield 40
Gloucester 22,40,38
Governor Holmes House 40
Grand Village 18,58,78
Great Sun 18
Green Leaves 41,70
Griffith-McComas House 41
Guest House 41,73,74
Guion, Isaac 22
gulf 14,24,118,120

—H—

Hannah, John 22
Happy Hunting Ground 18
Hawthorne 20
His Yankeeship 22
Holiday Inn 75,102
Holly Hedges 41
Holmes, David 23,25,40
Homewood 37,42
Homochitto River 19,79
Hope Farm 42,70,73
Hot Springs 25
hurricane 16
Hutchins, Anthony 19,22,42,43

—I—

Iberville 14,15
Illinois River 14
Independence 27,31
Indian Giving 19
Indigo 20,36,43,52
Inglewood 42
Ingraham, Joseph Holt 29
Institute Hall 42
Irishmen 22
Ivied Oaks 42

—J—

Jefferson College 25,29,54,58
Jersey Settlement 19,43,79
Jesuits 19
John Smith House 43
John Law 16,17
Johnson, William 22,26
Joliet, Louis 14

—K—

Kentucky Ark 24
Ker, David 25
King's Tavern 43
Kingston 19,43,79

—L—

La Salle 14,48,78

INDEX

La Ville de Rosalie aux Natchez 15
Lafitte, Jean 29
landlocked 23
Lansdowne 43,70
Liberty Hall 44
light house 26
Linden 44,46,70,73
Loess Bluff 65,77
Longwood 44,70
Lopez, Narciso 27
Louis XIV 14
loyalists 19,20
Lt. Governor 16

—M—

Magnolia Vale 120
Magnolia Hall 44,49,70,95,114
Marquette, Father Jaques 14
Marschalk, Andrew 24,52
Mason 23
massacre 17,18
Melmont 45
Melrose 45,73
Memorial Park 79,114
Memphis 14
Mercer House 45
Methodist 25,54,78,120
Mexican seed 24
mills 24
Minor, Stephen 22,76
Mississippi Herald 25
Mississippi Company 16-18
Mississippi Bubble 16
Mississippi Society 25
Mississippi Valley 14,23,49,63
Mississippi Gazette 25
Mississippian 16
Mistletoe 46,70
Mobile 15,16,20,23
Molasses Flats 46,95
Monette, John Wesley 29
Monmouth 27,45,46,70,73
monopoly 26,27
Montaigne 46,70
mounds 18,118
Mount Repose 47,73,70
Mount Locust 47
Murrell, John 23
Myrtle Terrace 47
Myrtle Bank 47
mystic clan 23

—N—

Napoleon 23
Nashville 22,29,80,
Natchez Under The Hill 9,22,24,80,88,90, 102
Natchez Trace 22,80,88,90,94,105,110,112
Natchez Institute 42
New England 29
New York 27,103,105
New Jersey 19,79
North Carolina 25
Nutt, Dr. Rush 24,44

—O—

Oakland 48,70
Ogden, Amos 19,79
Okolona 29
Old Glory 30,46

—P—

Paine Detected 25
Parsonage 48,70
Pearl River 15
Penicaut, Jean 14,15,18,78
Pennsylvania 19
Pensacola 14,17,20
Peter Christ House 49
Petit Gulf 24
pieces of eight 20
pirate 23,29
Pleasant Hill 45,49,74
Poindexter, George 25,27
Politics 22,25,28
Ponce de Leon 14
Ponchatrain 15
Pope, Piercy S. (Crazy) 25,53
Port Gibson 26
Postal 22,62
Postlewaite, Samuel 24
Prentiss Club 103
Presbyterian Manse 49
Presbyterian 25,39,45,49,54
Prince of the House of David 29
printing 24
Propinquity 49,74
Protestant 20

—Q—

quarantine 30
Quegles House 49

INDEX

Quitman, John Anthony 27,28,34,45,46,73,70

—R—

Ramada 36,75 95,103,107,114
Rattletrap 19
Ravenna 48
Ravennaside 74
Red River 17
revolutionaries 22
Richmond 50,70
Rio Grande de Florida 14
Rip Rap 51
Riverview 50
Roosevelt, Nicholas 24
Rosalie 15-18,22,30,40,48,50,70,77,78,88,94,122
Routhland 38,50,51,71

—S—

Salem Church 21
Sam Dale 23
Samuel Cockerell House 51
Santo Domingo 18
Saragossa 51
Sargent, Winthrop 22,25,40,38,51
Sargent's first laws 25
Savannah 86
Scotsman 16,19
Scott Home 52,120
Selma 52
Seven Years War 19,99
Sheraton 75,104,105,113,114
Ship Island 14
showboats 29
Shreve, Henry 24
Sicily Island 18,78
Smylie's Washington Academy 25
Spain 14,16,17,20,21,23,42,65
speaking bark 15
Springfield 52
St. Mary's Cathedral 51,79
Stanton Hall 39,42,37,52,74,70,96,102,105
Stars & Bars 30
steamboats 24
Stung Arm 18
Suns 17
Swayze, Richard & Samuel 19,21 79

—T—

temperance 27
Texada 14,15,27
Texas 14,15, 27
The Southwest, By A Yankee 29
theatre 100,105,107,108,110,112,119
Tillman House 53
Tobacco Panic 19
Tonty 48,78
tornado 26,35,77
Towers 53
Treaty of San Lorenzo 21
Trinity Episcopal Church 53
Tunicas 15
Turnipseed, Solomon 25
Twin Oaks 53,74,70

—U—

Unionist 28
United Mississippi Bank 60,68
University 25,60,118

—V—

Villa Gayoso 22
villains 23
Virginia 30,81
voting 25

—W—

Walking Johnson 22
Walsh House 54
Walworth Law Office 54
War of 1812 23
Watts, Margaret 22
welfare 16,31
Whigs 26,28
White Pillars 54
White Wings 54
White Turpin House 54
Wilkins House 54
Wilkinson, James 20,22
William Harris House 54
Williams, Robert 40,51,52
Williamson House 54
Willing, James 19,57

—X—Y—Z—

Yazoo 17,18,19
Yazoo Valley 17,18
Yellow Fever 41,25,30,37,46
Zion Chapel 54

AVA
BOOKS